Springer Theses

Recognizing Outstanding Ph.D. Research

Aims and Scope

The series "Springer Theses" brings together a selection of the very best Ph.D. theses from around the world and across the physical sciences. Nominated and endorsed by two recognized specialists, each published volume has been selected for its scientific excellence and the high impact of its contents for the pertinent field of research. For greater accessibility to non-specialists, the published versions include an extended introduction, as well as a foreword by the student's supervisor explaining the special relevance of the work for the field. As a whole, the series will provide a valuable resource both for newcomers to the research fields described, and for other scientists seeking detailed background information on special questions. Finally, it provides an accredited documentation of the valuable contributions made by today's younger generation of scientists.

Theses are accepted into the series by invited nomination only and must fulfill all of the following criteria

- They must be written in good English.
- The topic should fall within the confines of Chemistry, Physics, Earth Sciences, Engineering and related interdisciplinary fields such as Materials, Nanoscience, Chemical Engineering, Complex Systems and Biophysics.
- The work reported in the thesis must represent a significant scientific advance.
- If the thesis includes previously published material, permission to reproduce this must be gained from the respective copyright holder.
- They must have been examined and passed during the 12 months prior to nomination.
- Each thesis should include a foreword by the supervisor outlining the significance of its content.
- The theses should have a clearly defined structure including an introduction accessible to scientists not expert in that particular field.

More information about this series at http://www.springer.com/series/8790

Matías Vidal Navarro

Diffuse Radio Foregrounds

All-Sky Polarisation, and Anomalous
Microwave Emission

Doctoral Thesis accepted by
the University of Manchester, UK

 Springer

Author
Dr. Matías Vidal Navarro
Jodrell Bank Centre for Astrophysics
School of Physics and Astronomy
The University of Manchester
Manchester
UK

Supervisor
Prof. Clive Dickinson
Jodrell Bank Centre for Astrophysics
School of Physics and Astronomy
The University of Manchester
Manchester
UK

ISSN 2190-5053   ISSN 2190-5061   (electronic)
Springer Theses
ISBN 978-3-319-26262-8   ISBN 978-3-319-26263-5   (eBook)
DOI 10.1007/978-3-319-26263-5

Library of Congress Control Number: 2015955911

Printed on acid-free paper

This Springer imprint is published by SpringerNature
The registered company is Springer International Publishing AG Switzerland

Supervisor's Foreword

The thesis examines diffuse Galactic radio emission, focusing on the range $\sim$20–40 GHz. It is an important area of research because it is a major cosmological foreground for studies of the cosmic microwave background (CMB). It is also of interest for studying the properties and structure of the interstellar medium.

This substantial thesis work covers several topics and covers total intensity and polarisation. In particular, Vidal studied data from NASA's Wilkinson Microwave Anisotropy Probe (WMAP). These low signal-to-noise data must be corrected for polarisation bias. Vidal presents a new method to correct the data based on knowledge of the underlying angle of polarization. Using this novel method he sets upper limits on the polarization fraction of regions known to emit significant amounts of spinning dust emission. Vidal also studies the large-scale loops and filament that dominate the synchrotron sky. The dominant features are studied, including identification of several new features. A morphological link is established between one of the synchrotron spurs and the Fermi Gamma-ray bubbles. For the North Polar Spur, a model of an expanding shell in the vicinity of the Sun is tested, which appears to fit the data. Implications for CMB polarisation surveys are also discussed. Vidal also presents interferometric observations of the dark cloud LDN 1780 at 31 GHz. Vidal shows that the spinning dust hypothesis can explain the radio properties observed.

Manchester
November 2015

Prof. Clive Dickinson

Preface

During the past decades, the observation of the diffuse radio emission from the Galaxy has regained attention. In the 1950s, the first full-sky radio surveys were done at frequencies between 100 and 400 MHz. These observations, albeit at low angular resolution of $\sim 20°$, revealed the basic Galactic structure and the presence of 'spurs' of emission. Some of the physics of the interstellar medium (ISM) was also revealed by these surveys. The nonthermal synchrotron emission coming from individual supernova remnants (SNRs) and also from the diffuse ISM provides information about the cosmic rays propagation and magnetic fields on the Galaxy. The brightness temperature and spectra of HII regions that are seen in absorption at low frequencies can be used to study the density and temperature of the gas in star forming regions.

More recently, observations of the Cosmic Microwave Background (CMB) have reactivated the interest in the diffuse radio sky. The CMB is observed in the frequency range between 10 and 300 GHz and within this range there is strong emission from our Galaxy, both diffuse and from compact objects. As the brightest point sources can be masked out in the CMB analysis, the diffuse emission presents the greatest problem for CMB cosmology and its understanding and removal is critical. The importance of this foreground emission analysis was recently demonstrated with the publicized announcement by the BICEP2 team. They had claimed the detection of the B-mode polarisation pattern in the CMB, which would have major implications for cosmology, since gravitational wave B-modes are expected to be produced during the inflationary process when the Universe was less than 10^{-32} s old. Observations by the *Planck* satellite later showed that the polarised signal that the BICEP2 team reported as cosmological B-modes, was produced instead by Galactic dust. The search for B-modes continues and a number of ground-based experiments are measuring the polarised sky with even greater precision than BICEP2 and *Planck*. Balloon-borne and satellite experiments are also planned for the near future, so the necessity for an accurate quantification of the polarised foreground emission at $\sim$100 GHz is essential.

Polarised emission between 20 and 300 GHz is thought to be produced by two main mechanisms, namely synchrotron radiation from relativistic cosmic rays spiralling around magnetic field lines, and thermal radiation from dust grains that are aligned coherently by a local magnetic field. These two emission mechanisms have different spectral behaviour. Synchrotron has, at first order, a steep power-law spectrum where the emission decreases with frequency so that this emission dominates at lower frequencies. Thermal dust on the other hand has a "modified black body" spectrum, which peaks at $\sim$3 THz, with a rising spectrum in the frequency range we are interested in. The frequency where these two emission mechanisms show similar intensity is $\approx$70 GHz, and therefore is the frequency where the minimum polarised foreground emission is observed. This would correspond to the ideal frequency to observe CMB polarisation. However, most CMB experiments today use higher frequencies, between 150 and 300 GHz. The reason for this is that CMB observations are designed to avoid synchrotron contamination. Dust emission can be accurately measured at higher frequencies and then used to subtract the emission at the CMB range. In the case of synchrotron emission this is not possible at the moment as we do not have a polarisation map at low frequencies with the required accuracy. Moreover, the synchrotron spectral index is not uniform over the sky, as we will see in this work.

In total intensity observations, the task of CMB foreground analysis is even more difficult. Besides synchrotron and thermal dust, there is also free–free emission from plasma nearby massive stars, and the relatively recently discovered anomalous microwave emission (AME). AME is a new emission mechanism at ~ 30 GHz that is correlated with dust emission. It is thought to be originated by electric dipole emission from charged dust grains spinning at GHz frequencies. AME has been observed in many astrophysical environments and it is inherently diffuse. It is a major foreground contaminant for total intensity CMB observations. Until today, only upper limits of a few percentage points have been set up on its potential polarisation. Nevertheless, even if it is weakly polarized, it could complicate the CMB polarisation studies. AME is difficult to study due to its diffuse nature, being clearly detected by CMB experiments and telescopes at $\sim 1°$ angular resolution but hard to detect at higher angular resolution when interferometric arrays are used. This presents a problem for identification of the emitters and its physical properties so, at the moment, we only know some general properties of AME, like being associated with photon-dominated regions (PDRs). On the other hand, detailed theoretical models have been constructed that predict the spinning dust spectrum for different grain types and astrophysical environments. They present an opportunity to study the ISM, in particular the smallest dust grains, from a new window at GHz frequencies. Spinning dust emission depends on factors such as hydrogen density, gas temperature and ionisation fraction, so the detailed comparison of the models with good observations will allow us to study the ISM conditions in a variety of environments.

This thesis encompasses these topics: the study of the ISM in its relationship with CMB observations, both in total intensity and polarisation. The so-called "CMB foregrounds" science involves a diversity of astrophysical topics, including magnetic fields, cosmic rays, plasma and dust grains physics. This makes it a complex and interesting field, which due to the precision required for CMB cosmology, allows us to understand in great detail the physics of the Galactic ISM.

The motivation of this work is to study some aspects of the microwave sky, both in polarisation and total intensity. First, we focus on the polarised synchrotron radiation as observed by *WMAP*. The sky emission is seen to be dominated by large-scale filamentary structures which have not been extensively described in the literature. We identified and catalogued them and study some observational properties, such as spectral indices, polarisation angles distribution and polarisation fractions. A second topic studied is the anomalous microwave emission from molecular clouds. In particular, we study the LDN 1780 cloud using ancillary radio and infrared data as well as interferometric observations at 30 GHz. The idea is to identify the AME emitters, which are thought to be small dust grains.

The rest of the thesis is organised as follows. In Chap. 2, the *WMAP* polarisation data is described, along with the further processing that we perform to the data. The polarisation amplitude P, being a definite positive quantity suffers noise bias. We describe the methods that exist to correct for this effect and present the corrected *WMAP* polarisation intensity maps that we use in the rest of this thesis. We also describe an application of the bias-correction technique to the quantification of the polarisation fraction of two AME dominated regions. In Chap. 3, we study the polarised diffuse emission using *WMAP* data. The work is focused on the characterisation of the polarised filamentary emission visible in the maps. The geometry of these features is studied as well as spectral indices and polarisation fractions. We also include an interpretation of the large-scale features and their possible connection to the local ISM. Chapter 4 is based on the Q/U Imaging Experiment (QUIET), a polarisation CMB experiment which observed regions of the sky at 43 and 95 GHz. We describe the instrument, observations and map-making process focused on the data from two regions observed on the Galactic plane. A comparison of the maps with *WMAP* data is shown and some basic analysis of the diffuse polarised emission on the Galactic regions. In Chap. 5 we use new interferometric data at 31 GHz to study the AME from the LDN 1780 cloud. The connection between the microwave emission and dust is studied using IR data that traces different dust grains. The parameter space of spinning dust models is explored and we compare with the observed physical conditions of the cloud. Chapter 6 summarises the main results of this work and also describes the future work.

Dr. Matías Vidal Navarro

Acknowledgments

When I first met Prof. Richard Davis, he said to me that a Ph.D. is to show that the applicant is able to write a thesis. This sounded weird to me at that time, as I thought that the most difficult part is doing the research and the writing was just a necessary ending for the process, without much complication. Years later, while I was struggling to finish this writing of mine, those words kept coming constantly to my mind. The writing of my thesis, a comparatively short but very intense period of my Ph.D., was the conclusion of almost 4 years of work and learning. A number of people were very important for me to achieve my degree, so I would like to mention them here.

I would like to thank my supervisor, Dr. Clive Dickinson, for constantly pushing me to my best. I will keep with me his all his advice regarding all the aspects of my career, not only the purely scientific. He gave me enough freedom to organise my time and direct my investigations, which has helped me to become more independent in my research. His ability to see clearly the 'big picture' of the situations helped me to regain focus several times.

I am also very grateful for being able to work closely with Prof. Rod Davies. I felt that he transferred to me some of his deep love for this science and his appreciation for the natural world. This helped me to gain perspective during stressful moments by realising that we are fortunate for being able to spend our time trying to understand Nature.

I am also happy for having worked with Dr. Paddy Leahy. His deep understanding of complicated issues allowed him to answer, in concrete and simple terms, my constant questions. He is also very generous with his time, so besides being an excellent scientist he is also one of the best teachers I have met.

I would also thank my two Ph.D. examiners, Dr. Tony Banday and Dr. Michael Keith, for a careful reading of my thesis and for useful comments that greatly improved the quality of the final version.

A number of people also helped me on different topics during my Ph.D.; among them I would like to thank Prof. Hans Kristian Eriksen, Dr. Ingunn Wehus, Dr. Joe Zunts, Dr. Adam Avison, Dr. Bob Watson and Prof. Richard Battye.

My Ph.D. in Manchester was a time when I think I grew up in many aspects, thanks to the great people I encounter within and outside work. I would like to thank in particular Florence, for her tireless support and comprehension. My good friends and Mancunian family, Cristobal, Pedro, Melis, Alicia, Sotiris, Wamz, Adam, Citlalitl, Sinaman, Silvia, Stefania, Tania, Carolina, Viviana, Hugo, Felix, Beatriz, Valentina, Patrick, and Marta, for the endless conversations and great moments, which, at the end, will remain the longest dearly in my mind.

Finally but most importantly, to my family, for their love and constant support from the distance, which was most important as a foreigner student.

Contents

About the Author

Matías Vidal Navarro was born in December 1984 in Santiago, Chile. He obtained an MSc in Astrophysics in September 2009 at the Universidad de Chile. Then, in March 2010, he started a Ph.D. at the University of Manchester under the supervision of Dr. Clive Dickinson. This thesis presents the results of the research work done during his Ph.D.

Abbreviations

ΛCDM	Lambda-Cold Dark Matter
AME	Anomalous Microwave Emission
BICEP	Background Imaging of Cosmic Extragalactic Polarization
CARMA	Combined Array for Research in Millimeter-wave Astronomy
CBI	Cosmic Microwave Imager
CES	Constant Elevation Scan
CMB	Cosmic Microwave Background
COBE	Cosmic Background Explorer
CR	Cosmic Rays
DRAO	Dominion Radio Astrophysical Observatory
FWHM	Full-Width at Half Maximum
HEMT	High Electron Mobility Transistors
ILC	Internal Linear Combination
IR	Infrared
ISO	Infrared Space Observatory
ISM	Interstellar Medium
ISRF	Interstellar Radiation Field
LDN	Lynds Dark Nebula
MCMC	Monte Carlo Markov Chains
MEM	Maximum Entropy Method
NPS	North Polar Spur
PAH	Polycyclic Aromatic Hydrocarbons
QUIET	Q/U Imaging ExperimenT
RM	Rotation Measure
RRL	Radio Recombination Line
SED	Spectral Energy Distribution
SN	Supernova
SNR	Signal-to-Noise Ratio
TOD	Time Ordered Data

UV	Ultraviolet
VSG	Very Small Grains
WMAP	Wilkinson Microwave Anisotropy Probe

List of Figures

List of Tables

Chapter 1
Introduction

As a result of great theoretical and observational work during the last century, our understanding of the Universe has change dramatically. Today, we have a concordance cosmological model, based on an isotropic and homogeneous universe under the framework of general relativity. The universe is expanding and the rate of expansion is observed to be increasing at the present time. The baryonic matter constitutes less than 5 % of the total energy density of the universe. A dark matter component is invoked to account for the "missing mass", which is only revealed by its gravitational interaction with ordinary (baryonic) matter and photons. Dark matter constitutes about 27 % of the energy density. An even more mysterious component, the dark energy, is necessary to explain the accelerated expansion of the universe. This component is the largest, comprising more than 68 % of the total (Planck Collaboration et al. 2013a). This scenario is referred as the ΛCDM model and at the moment is the one that better describes all the observational data available (see e.g. the review by Frieman et al. 2008). An early period of accelerated expansion, where the universe increases its volume by a factor $\sim e^{60}$ is necessary (Ryden 2003). The *inflation* solves a number of observational properties of the universe, like its statistical homogeneity and isotropy, and naturally predicts the formation of the large scale structures seen in the universe (see e.g. Liddle and Lyth 2000; Linde 2008; Beringer et al. 2012).

The observation and very precise measurements of the Cosmic Microwave Background (CMB) has played a key role in defining our modern cosmological view. The CMB is the thermal relic radiation from a moment when the universe was about 1000 times more compact and hotter than today. It is a snapshot of the early universe, in which are recorded the "initial conditions" that will later evolve to form the structures we see today. Moreover, it can give us a glimpse of the conditions when the universe was less than 10^{-32} of a second old. A spectrum of gravity waves, expected to be produced in some inflationary models, may yield a detectable polarisation field on the CMB, the *B-modes*. The detection of this pattern would confirm the inflationary paradigm and will give insights into the extremely high energy scale of the early universe.

© Springer International Publishing Switzerland 2016
M. Vidal Navarro, *Diffuse Radio Foregrounds*, Springer Theses,
DOI 10.1007/978-3-319-26263-5_1

The CMB is observed at GHz frequencies. In this frequency range, there is also emission from different astrophysical objects, in particular diffuse gas and plasma from our Galaxy. In this context, the diffuse Galactic emission is a foreground contamination to the cosmological signal of the CMB and important efforts are made for its characterisation and removal. The emission from B-modes is expected to be orders of magnitude smaller than the CMB temperature, and a very precise characterisation of the foreground emission is necessary. A number of experiments are involved in this search and the data that they are providing are increasing greatly our understanding in cosmology and also the astrophysics of the foregrounds. The CMB investigations have revitalised the study of the diffuse Galactic emission, which started more than 60 years ago with the first radio astronomy investigations. In the rest of this chapter I first describe the CMB and then, the other emission mechanisms that dominate the GHz sky, both in total intensity and polarisation.

1.1   CMB

The emission from the Cosmic Microwave Background (CMB) was serendipitously discovered as a 3.5 K excess at 4.08 GHz, which was *"within the limits of our observations, isotropic, unpolarized, and free from seasonal variations"* (Penzias and Wilson 1965). It was clear that this radiation could be the remnant from an earlier and hotter stage of the universe and this idea was published in a companion paper by Dicke et al. (1965), who, at the time, was constructing a radiometer to detect this radiation. The existence of the CMB, as a consequence of a hot origin of the universe, and its temperature at the present time (to be 5 K) were first predicted by Alpher and Herman (1948, 1949). The discovery of the CMB brought great attention to cosmology and important theoretical and observational work has been done since then. Crucial observations came in 1990 with the Cosmic Background Explorer (*COBE*) satellite, which measured the spectrum of the CMB with unprecedented precision over the full sky, up to angular scales of $\sim 7°$ (see e.g. Boggess et al. 1992, for a description of the mission). The onboard Far-Infrared Absolute Spectrophotometer (FIRAS) instrument measured the CMB intensity between 30 and 2910 GHz and the data are fitted with a nearly perfect black-body spectrum with temperature 2.7260 ± 0.0013 K (Fixsen 2009). The deviations from the blackbody spectra were constrained to be less than 50 parts per million. These measurements are one of the observational pillars of the current standard model of cosmology.

The CMB is the relic of a period when the universe was in thermal equilibrium. A high temperature photon-baryon plasma is tightly coupled due to Compton scattering and electric interactions. The expansion of the universe cools down the plasma and when the temperature has reached ~ 3000 K, most photons have an energy lower than 13.2 eV, necessary for the ionisation of hydrogen. At this period, neutral hydrogen atoms appear for the first time, as electrons start being captured by the protons. The redshift at which this *recombination* occurs is $z \sim 1100$, with little dependence on the details of the model (Scott and Smoot 2004). Before this period, the mean free path of the photons was very small, and the universe was opaque to radiation. The CMB acts as an effective barrier, limiting how far back in time we can learn from the universe using photons.

Fantastically, new optical observations of distant galaxies allow the measurement of the CMB temperature over cosmic time. Noterdaeme et al. (2011) measured $T_{CMB} = 10.5^{+0.8}_{-0.6}$ K at $z = 2.68$, i.e., when the Universe had only 20% of its present age. This measurement is consistent with the current standard cosmological model.

The *COBE* satellite also brought another outstanding discovery. A second instrument, the Differential Microwave Radiometer (DMR), showed statistically the existence of spatial variations in the CMB temperature, the *anisotropy* with an amplitude of $\Delta T / T \approx 1 \times 10^{-5}$ (Smoot et al. 1992).

1.1.1 CMB Anisotropies

The temperature of the CMB is not uniform across the sky. There are small fluctuations that can be conveniently expressed as deviations over the mean CMB temperature using spherical harmonics,

$$\frac{\Delta T}{T} = \sum_{l,m} a_{\ell m} Y_{\ell m}(\theta, \phi), \tag{1.1}$$

where the multipole value $\ell \sim \pi/\theta$. In this decomposition, the mean temperature of the CMB (the monopole) corresponds to $\ell = 0$.

If the temperature fluctuations are Gaussian and isotropic, then all the statistical information of the anisotropies is contained in the CMB power spectrum,

$$C_\ell^{TT} = \langle a_{\ell m}^{T*} a_{\ell' m'}^{T} \rangle \tag{1.2}$$

The largest temperature anisotropy that we observe is the dipole ($\ell = 1$). It is originated by the Doppler shift of the measured temperature, as the solar system moves with respect to an inertial frame defined by the CMB. Its value was measured by *COBE* to be 3.381 ± 0.007 mK, which corresponds to a velocity of 372 ± 1 km s^{-1} in the direction $(l, b) = (264°.14 \pm 0°.15, 48°.26 \pm 0°.15)$ (Fixsen and Mather 2002).

The rest of the fluctuations observed in the CMB can be classified as *primary anisotropies*, which are imprinted during the recombination period and *secondary anisotropies*, that are produced due to scattering of the CMB photons along the line-of-sight, during their travel from the last scattering surface to our telescopes.

The primary anisotropy is caused by variations in the matter density before recombination, when baryons and photons are tightly coupled. Small ($\mathcal{O}(10^{-5})$) gravitational perturbations compress the plasma. This increases the photon pressure which acts as a restoring force, producing *acoustic oscillations* in the plasma. When photons decoupled from matter at recombination, the phases of the oscillations were frozen-in, so the more compressed (diluted) regions appear as a hotter (colder) deviation from the mean CMB temperature.

The size of the horizon at the period of recombination determines the largest scale in which there is causal contact. Perturbations over larger scales have a period longer than the age of the Universe at that time. The fluctuations observed at $\ell < 100$ represent large scale variations in the gravitational potential during recombination, which through gravitational redshift are imprinted on the CMB. This anisotropy is observed as a plateau in the power spectrum and it is referred as the Sachs-Wolfe effect (Sachs and Wolfe 1967). Additionally, the gravitational redshift that affects the CMB photons in their travel from the last scattering surface towards us is called the integrated Sachs-Wolfe effect, so it is not part of the primordial CMB fluctuations.

On smaller (sub horizon) scales, the *acoustic peaks* are the dominant feature in the power spectrum between $100 \lesssim \ell \lesssim 1000$. The peaks represent the scales of maximum compression and expansion of the plasma. The valleys in between the peaks are non-zero due to Doppler shifting of the emission, as they correspond to the maximum in the velocity of the plasma.

At smaller scales ($\ell \gtrsim 1000$), the power spectrum samples scales that are smaller than the thickness of the last scattering surface (the recombination processes is not instantaneous). This produces a dampening of the spectrum, due to the superposition of different temperatures which washes-out the small scale details. At the high ℓ range, the spectrum is also affected by secondary fluctuations. Scattering of the CMB photons with hot electrons in the halo of galaxy clusters generates spectral fluctuations in the CMB radiation. This is known as the Sunyaev-Zeldovich (SZ) effect (Sunyaev and Zeldovich 1970). Another secondary anisotropy is produced by gravitational lensing of the CMB photons by structures in the local universe, which subtly broader the peaks.

1.1.2 Observational Status of the Temperature Power Spectrum

The exact location and relative amplitude of the features in the power spectrum are defined by the parameters of the cosmological model. The measurement of the temperature fluctuations, since its statistical discovery by *COBE*, has contributed to lead us into an era of "precision cosmology". The combination of complementary data from ground-based, balloon and satellite experiments has tightly constrained the power spectrum up to angular scales of 5 arcmin ($\ell \sim 2500$). Among these experiments, the observations by the Wilkinson Microwave Anisotropy Probe (*WMAP*, Bennett et al. 2013 and references therein) between 2001 and 2010 stand out. Observing in five frequency bands between 23 and 94 GHz, it was able to measure the acoustic peaks up to $\ell \sim 800$, constraining the parameter-space of the cosmological parameters by a factor of 68,000, relative to pre-*WMAP* measurements

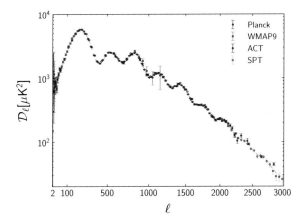

Fig. 1.1 CMB temperature power spectrum including data from *Planck*, *WMAP*, ACT and SPT. The vertical axis represents the value $\ell(\ell+1)C_\ell/2\pi$ as a function of the multipole ℓ. The data is fitted by a six-parameter ΛCDM model. The peaks and troughs are the imprint of the acoustic oscillations. The decrease of the power from $\ell \sim 1000$ is due to silk dampening. At the largest angular scales ($\ell \lesssim 50$), the Sachs-Wolfe plateau dominates the power but in this figure, it is not clear due to the logarithm scale of the x-axis. This figure is reproduced from Planck Collaboration et al. (2013a)

(Hinshaw et al. 2013). In 2009, the *Planck* spacecraft was launched to take the lead in the full-sky measurements. It uses nine frequency bands and, with low noise receivers, precisely measures the anisotropy up to $\ell \sim 2500$ (Planck Collaboration et al. 2013b).

The telescope size is a major constraint in space missions. This is not the case for ground-based experiments and they have proved to be fundamental in the measurement of the anisotropy at high-ℓ. This area of research has been dominated during the last few years by the South Pole Telescope (SPT, Carlstrom et al. 2011) and the Atacama Cosmology Telescope (ACT, Swetz et al. 2011). Both telescopes measure the CMB at arcmin scales ($\ell \sim 10,000$) from extremely dry locations (the South Pole and the Atacama Desert in Chile). The measurement of the damping tail of the CMB and the quantification of secondary anisotropy effects (SZ effect, CMB lensing) are the key targets of these projects. Both telescopes have remarkably detected up to the ninth acoustic peak of the power spectrum (Story et al. 2012; Das et al. 2013). The combination of these data with the lower-ℓ spectra measured by *WMAP* and *Planck* is a strong confirmation of the ΛCDM model. In Fig. 1.1 we show the temperature power spectrum state as of March 2013.

1.1.3 CMB Polarisation

Not all the cosmological information that we can obtain from the CMB is encoded in the temperature fluctuations. The primordial density perturbations leave an imprint

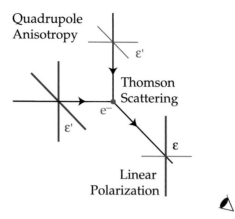

Fig. 1.2 Modified diagram from Hu and White (1997) showing the emergence of polarisation due to Thompson scattering by an electron embedded in a quadrupolar radiation field. The radiation that arrives from the *left* to the electron would produce a vertical polarisation for the observer, as the perpendicular component is along the line-of-sight and not visible. Analogously, the radiation incoming to the electron from the *top* yields a horizontal polarisation after the scattering. Because the intensity on the incident radiation is different (the radiation coming from the *left* is *hotter* than the one arriving from the *top*) a net linear vertical polarisation is measured by the observer

on the last scattering surface also as a polarisation pattern. The polarisation anisotropy of the CMB can only be generated via scattering, both at the recombination surface and in posterior interactions of the CMB radiation with free electrons along a line-of-sight. The value of the CMB polarisation depends on the angular scale, and an average value is $\sim 10\%$ on angular scales of a few arcmin.

The polarisation anisotropy also arises as a consequence of perturbations in the plasma at the moment of recombination. A quadrupolar component of the radiation field, i.e. changes in the intensity of the radiation from perpendicular directions, will produce a net polarisation. A single electron embedded in such anisotropic field would Thompson-scatter light with a non-zero polarisation intensity. This is shown as a sketch in Fig. 1.2.

Due to the orthogonality of the spherical harmonics, $Y_{\ell m}(\theta, \phi)$, the quadrupole component ($\ell = 2$) is the only ℓ mode that can generate polarisation (Hu and White 1997). There are three geometrical components of the quadrupole, namely *scalar*, *vector* and *tensor* modes, that correspond to the allowed values for m, which are $m = (0, \pm 1, \pm 2)$. *Scalar* fluctuations generate velocity gradients in the plasma, as the matter falls and bounces back from the potential wells at the last scattering surface. The velocities of the plasma are out of phase with the density distribution. The scalar fluctuations are also the origin of most of the temperature anisotropy. *Vector* fluctuations corresponds to vorticities in the plasma at the last scattering surface. They are expected to be extremely low as the early inflation vanishes any primordial vorticity. *Tensor* fluctuations, that can be interpreted as gravitational waves, compress and stretch the space in perpendicular directions, compressing and relaxing the plasma and generating a quadrupole pattern in the temperature fluctuations.

1.1.3.1 Polarisation Patterns

From an observational point of view, polarised electromagnetic radiation can be described using the Stokes parameters. The electric field, **E**, of a electromagnetic wave propagating in the z-direction can be expressed as:

$$\mathbf{E} = a_x(t) \cos[\omega t - \theta_x(t)]\hat{\mathbf{x}} + a_y(t) \cos[\omega t - \theta_y(t)]\hat{\mathbf{y}},\tag{1.3}$$

where ω is the frequency of the radiation and θ_x and θ_y are the phase angles of the oscillation. The Stokes parameters are defined as time averages of the field:

$$
\begin{aligned}
I &\equiv \langle a_x^2 \rangle + \langle a_y^2 \rangle \\
Q &\equiv \langle a_x^2 \rangle - \langle a_y^2 \rangle \\
U &\equiv \langle 2a_x a_x \cos(\theta_x - \theta_y) \rangle \\
V &\equiv \langle 2a_x a_x \sin(\theta_x - \theta_y) \rangle.
\end{aligned}\tag{1.4}
$$

I corresponds to the total intensity of the field and, in the case of the CMB blackbody emission, it represents the temperature of the plasma. Q and U represent the linear polarisation while V is the as circular polarisation. Thompson scattering does not produce circular polarisation, so in CMB studies $V = 0$. The measuring of V in CMB experiments can be used to detect systematic errors or additional foreground contamination. Other useful quantities are the polarisation amplitude P and the polarisation angle χ, which is parallel to the direction of the electric field:

$$
\begin{aligned}
P &= \sqrt{Q^2 + U^2} \\
\chi &= \frac{1}{2}\arctan\left(\frac{U}{Q}\right).
\end{aligned}\tag{1.5}
$$

The polarised signal can be decomposed into spherical harmonics, analogously to the temperature decomposition,

$$(Q \pm iU)(\theta, \phi) = \sum_{\ell,m} a_{\ell,m}^{\pm 2} Y_{\ell,m}^{\pm 2}(\theta, \phi).\tag{1.6}$$

Furthermore, this polarisation pattern in harmonic space can be separated into a curl-free "electric" E-mode and a divergence-free "magnetic" B-mode. This representation is more useful in CMB analysis because they describe the polarisation with respect to the wave itself and not to an arbitrary coordinate system (like N-S or E-W). Therefore, if the polarisation is parallel or orthogonal to the direction of the radiation, it is an E-mode, if the polarisation is orientated at 45°, corresponds to a B-mode. The E and B-mode are defined as follows (Zaldarriaga and Seljak 1997),

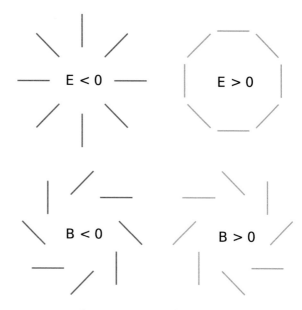

$$E(\theta, \phi) = \sum_{\ell,m} a^E_{\ell,m} Y_{\ell,m}(\theta, \phi) \equiv \sum \left(-\frac{1}{2}(a^{(2)}_{\ell,m} + a^{(-2)}_{\ell,m}) \right) Y_{\ell,m}(\theta, \phi) \qquad (1.7)$$

$$B(\theta, \phi) = \sum_{\ell,m} a^B_{\ell,m} Y_{\ell,m}(\theta, \phi) \equiv \sum \left(-\frac{1}{2i}(a^{(2)}_{\ell,m} - a^{(-2)}_{\ell,m}) \right) Y_{\ell,m}(\theta, \phi). \qquad (1.8)$$

Figure 1.3 shows examples of E- and B-mode.

By analogy with Eq. 1.2 the polarisation power spectra of the E and B-mode are defined as

$$C^{EE}_\ell = \langle a^{E*}_{\ell m} a^E_{\ell'm'} \rangle$$
$$C^{BB}_\ell = \langle a^{B*}_{\ell m} a^B_{\ell'm'} \rangle$$
$$C^{TE}_\ell = \langle a^{T*}_{\ell m} a^E_{\ell'm'} \rangle \qquad (1.9)$$

where the third spectrum, C^{TE}_ℓ, is the cross correlation between the temperature and E-modes anisotropy. Parity invariance demands that the cross correlations C^{TB}_ℓ and C^{EB}_ℓ are zero (Hu and Dodelson 2002).

Another key characteristic of the E and B decomposition is that E-modes can only be produced by scalar fluctuations. B-modes, on the other hand, can be generated by scalar and tensor fluctuations. This way, gravitational waves perturbations could leave an imprint on the B-mode spectrum. The amplitude of the tensor fluctuations is usually characterised by the *tensor-to-scalar ratio, r*. This value is expected to be higher in inflationary models that have higher energy and different models would lead to different values for r. For instance, models where the inflationary potential is described as $\lambda\phi^4$, predict $r = 0.32$, while other models can lead to arbitrarily small values for r (Scott and Smoot 2004). An additional component to the B-mode spectrum is produced through gravitational lensing by galaxies of the E-mode signal. This effect is therefore noticeable at small angular scales ($\ell \gtrsim 500$) and has recently been detected using a cross-correlation with a lensing template (Hanson et al. 2013).

1.1.4 Observational Status of the CMB Polarisation

The E-mode spectrum was observed for the first time in 2005 by the Degree Angular Scale Interferometer (DASI, Leitch et al. 2005). Since them, C_ℓ^{EE} and also C_ℓ^{TE} have been measured by many experiments (Montroy et al. 2006; Sievers et al. 2007; Wu et al. 2007; Bischoff et al. 2008; Brown et al. 2009; Larson et al. 2011; QUIET Collaboration et al. 2011). The search for B-modes has been more elusive. Even in the inflationary models with the largest values of r, the B-mode spectrum is expected to have at least one order of magnitude less power than the E-mode spectrum. At the moment, there are only upper limits by a number of experiments. In Fig. 1.4 is shown the state of the polarisation spectra C_ℓ^{EE} and C_ℓ^{BB} as of December 2012. The current best upper limit for the tensor-to-scalar ratio, coming from CMB polarisation only measurements is $r < 0.70$ at the 95 % confidence level (BICEP1 Collaboration et al. 2013), while the best constraint including the temperature power spectra is $r < 0.11$, also at the 2σ level (Planck Collaboration et al. 2013c). A number of ground based and balloon-borne experiments are aiming to reach the $r = 0.01$ level. Some of them are already in operation or being constructed (Essinger-Hileman et al. 2010; Niemack et al. 2010; Ogburn IV et al. 2010; Eimer et al. 2012; Oxley et al. 2004; Sheehy et al. 2010; Benford et al. 2010; Fraisse et al. 2013; Arnold et al. 2010; Austermann et al. 2012). The polarisation results from the *Planck* mission are expected to be available in 2014.

1.2 Emission Mechanisms at GHz Frequencies

The radio sky in the 1–100 GHz frequency range is very interesting as different emission mechanisms contribute to the total brightness. Different regions of the sky are dominated by distinct types of radiation and only inside this frequency range the CMB temperature anisotropy is larger than the various foreground emission in some regions of the sky.

The radiation from astrophysical objects is produced by charged particles undergoing an acceleration or by quantum transitions in atoms or molecules. A detailed treatment of radiative processes can be found in Rybicki and Lightman (1979) while radio emission mechanisms are covered by Wilson et al. (2009), Longair (1994). In this section we describe the emission mechanisms that are important in the diffuse microwave sky, which are:

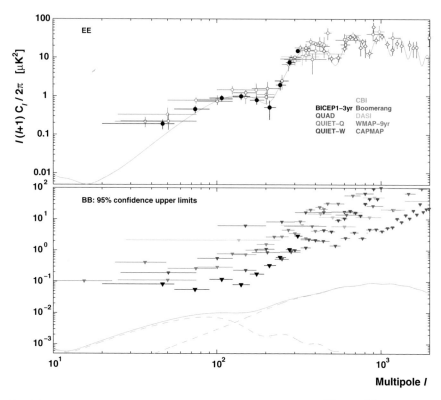

Fig. 1.4 Current observational status of the polarisation power spectra C_ℓ^{EE} and C_ℓ^{TE} as of October 2013. The EE spectrum on *top* has been measured by various experiments. The upper limits on the B-mode spectra are shown at the *bottom*. Here, the *dash line* that peaks around $\ell = 90$ corresponds to the spectra produced by gravitational-waves if the tensor-to-scalar ratio has a value of $r = 0.1$. The second peak at $\ell \sim 1000$ is due to gravitational lensing and is not related to the primordial gravitational wave background. This figure is reproduced from BICEP1 Collaboration et al. (2013)

- Synchrotron radiation
- Free-free emission (Bremsstrahlung)
- Thermal dust emission
- Anomalous microwave emission

The average contribution over a large fraction ($\sim 80\%$) of the sky, away from the Galactic plane, of these different emissions, as measured by the *WMAP* satellite (Bennett et al. 2013), is shown in Fig. 1.5. In this figure we can see that the minimum foreground contamination in temperature on large angular scales, occurs at a frequency $\sim 70\,\text{GHz}$. The CMB anisotropy dominates the spectrum between $\sim 40\,\text{GHz}$ and $\sim 120\,\text{GHz}$, where the thermal dust emission begins to dominate the emission in the sky.

Fig. 1.5 Spectra of different emission components in the GHz range as measured by *WMAP*. The *vertical coloured regions* represent the frequency range of each of the *WMAP* bands. This figure is reproduced from Bennett et al. (2013)

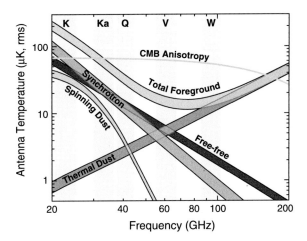

1.2.1 Synchrotron Radiation

Charged particles accelerated by a magnetic field will radiate. The diffuse Galactic synchrotron emission is produced by cosmic rays, relativistic particles (principally electrons) with very high energies spiralling around Galactic magnetic field lines. These particles are accelerated to relativistic speeds in very energetic environments, such as shock-waves from supernovae explosions.

The synchrotron intensity depends on the electron density (n_e), the energy distributions of these electrons and the strength of the magnetic field. The energy distribution of cosmic rays (CR) is observed to follow a power-law, $N(E) \propto E^{-p}$ for energies $E > 10\,\text{GeV}$ (e.g. Adriani et al. 2011; Ackermann 2012). This distribution is fairly smooth, both spectral and spatially over the sky (Strong et al. 2004, 2007). Typical values measured for the spectral index p of the distribution are close to $p = 3$, and therefore, this value is usually adopted in synchrotron and magnetic field modelling (e.g. Miville-Deschênes et al. 2008; Jaffe et al. 2010).

The intensity of the synchrotron radiation, for a power-law distribution of energy, at a frequency ν is (e.g. Miville-Deschênes et al. 2008),

$$I_{\text{sync}}(\nu) = \varepsilon_{\text{sync}}(\nu) \int_z n_e B_\perp^{(1+p)/2} \mathrm{d}z, \qquad (1.10)$$

where the integral is along the line-of-sight z and $B_\perp = \sqrt{B_x^2 + B_y^2}$ is the component of the magnetic field on the plane of the sky. The emissivity corresponds to a power-law:

$$\varepsilon_{\text{sync}}(\nu) = \varepsilon_0 \nu^{-(p-1)/2}. \qquad (1.11)$$

This intensity can be converted to a brightness temperature using the Rayleigh-Jeans law,

$$T_{\text{sync}}(\nu) = \frac{c^2 I_{\text{sync}}(\nu)}{2 k \nu^2}. \tag{1.12}$$

Therefore, the brightness temperature of synchrotron emission in a particular line-of-sight can be written as

$$T(\nu) = T(\nu_0) \left(\frac{\nu}{\nu_0}\right)^{\beta_s} \tag{1.13}$$

with $\beta_s = -(p+3)/2$.

The spectral index can vary spatially across the sky (e.g. Larson et al. 1987; Reich and Reich 1988) and also with frequency (see e.g. Banday and Wolfendale 1991; Giardino et al. 2002; Platania et al. 2003). This is attributed to CR ageing; old electrons have lost more energy due to synchrotron emission. The losses of energy of the cosmic rays through synchrotron radiation are larger at higher energies ($\propto E^2$), so over time, the synchrotron spectral index becomes steeper ("more negative"). β_s is also expected to be steeper with higher Galactic latitude (Strong et al. 2007). This is because most cosmic rays are produced in super nova (SN) explosions, close to the Galactic plane. As the electrons diffuse away from the plane, to higher Galactic latitudes, they will lose some of their energy, changing β_s Averaged values for the measured spectral indices varies from $\beta = -2.55$ between 45 and 408 MHz (Guzmán et al. 2011), $\beta = -2.71$ between 408 MHz and 2.3 GHz (Platania et al. 2003; Giardino et al. 2002) and $\beta = -3.01$ between 23 and 33 GHz (Davies et al. 2006; Dunkley et al. 2009). We note that, at frequencies of tens of GHz, the sky emission has different components, which differ in their spectrum and in their spatial distribution. This makes the determination of the synchrotron spectral index a complicated problem, which requires the use of some component separation technique.

The 408 MHz all sky survey (Haslam et al. 1982) is a traditionally used template for Galactic synchrotron radiation because, at the frequency of the survey, synchrotron is expected to dominate. Figure 1.6 shows this map in Galactic coordinates. Most of the emission in this maps originates at the Galactic plane. At high Galactic latitudes, the most important feature is the North Polar Spur (NPS), which runs roughly perpendicular to the Galactic plane, towards positive latitudes. We will discuss this feature in detail in Chap. 3.

1.2.2 Synchrotron Polarisation

Synchrotron emission is intrinsically linearly polarised. The magnetic field lines provide a privileged direction for the motion of the emitting particles, so the emission is expected to be polarised. Given a uniform magnetic field perpendicular to the line-of-sight, the fractional polarisation is expected to be $\Pi = (p+1)/(p+7/3)$ (Rybicki

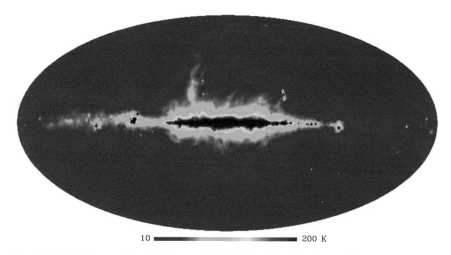

10 ▬▬▬▬▬▬▬▬▬▬ 200 K

Fig. 1.6 408 MHz map from Haslam et al. (1982) showing the sky at a frequency where the emission is dominated by diffuse synchrotron emission. The *linear colour* scale has been truncated at the 200 K value to highlight the more diffuse features at high Galactic latitudes. A number of extragalactic sources are visible in the map, including the Cen A galaxy and the *magellanic clouds*

and Lightman 1979). For a typical value of cosmic ray spectral index ($p \approx 3$) the polarisation faction can be as high as 75 %. The observed value is smaller than this due to the superposition effects along the line-of-sight, beam averaging and intrinsic misalignment of the magnetic field (see e.g. Macellari et al. 2011; Kogut et al. 2007).

The necessity of a good synchrotron template at higher frequencies for CMB foreground estimation has motivated a number of experiments that map the sky both in total intensity and polarisation. Among these are the C-Band All Sky Survey (CBASS) at 5 GHz (King et al. 2013), the S-band Polarization All Sky Survey (S-PASS) at 2.3 GHz (Carretti 2010).

In Chap. 3 of this thesis, we study the observational properties of the polarised emission in the 23–40 GHz range, which is dominated by synchrotron emission. In that Chapter, we will discuss in more detail the properties of polarised synchrotron.

1.2.3 Free-Free Emission

Bremsstrahlung or free-free emission is produced by the acceleration of a charged particle (e.g. electron) being deflected by another charged particle (e.g. proton). It is worth mentioning that the net electric dipole in the interaction must be non-zero to have radiation, so an encounter between two particles with the same charge and mass (e.g. two electrons) will not produce radiation. The acceleration of the particles during the encounter is not constant, so the emitted photons have a range of wavelengths (Rybicki and Lightman 1979).

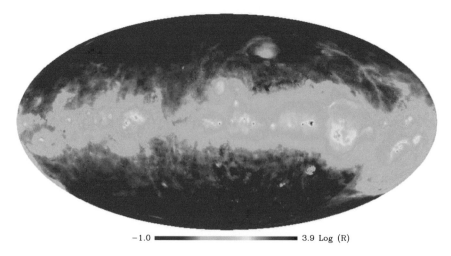

-1.0 ▬▬▬▬▬▬▬▬▬▬▬▬▬▬▬▬▬ 3.9 Log (R)

Fig. 1.7 Hα map without any correction for extinction from Finkbeiner (2003). This map is used to trace the free-free emission in regions where the dust absorption is not dominant (away from the Galactic plane). The units are Rayleighs and the *colour scale* is logarithmic to highlight the fainter areas. Most of the emission comes from HII regions around star forming clouds

The free-free spectral index β_{ff} is a slowly varying function of frequency and electron temperature (Bennett et al. 1992),

$$\beta_{ff} = -\left(2 + \frac{1}{10.48 + 1.5\ln(T_e/8000\,\mathrm{K}) - \ln(\nu_{\mathrm{GHz}})}\right), \qquad (1.14)$$

which at GHz frequencies gives an average $\beta_{ff} \approx -2.1$. The optical Hα line is a good tracer of free-free emission as it also depends on electron-ion interactions. The radio and Hα emission are both proportional to the *Emission Measure* (EM∝ $\int n_e^2 dl$, with n_e the electron density). Observationally the estimation of free-free by using the Hα optical line requires a correction for dust absorption and this is the major source of uncertainty, mostly in the dense regions of the Galaxy (see e.g. Dickinson et al. 2003). Figure 1.7 shows a full-sky Hα map from Finkbeiner (2003) that has not been corrected for dust extinction, so it only traces the free-free emission on regions away from the Galactic plane, where the dust extinction is less important.

Another way of estimating the radio contribution of free-free is by using hydrogen radio recombination lines (RRL). RRL data are used to measure the emission measure (EM = $\int n_e^2\, dl$) along the line-of-sight without the absorption limitation that affects the Hα line. If the electron temperature is known, the free-free emission can be calculated from the RRLs (see e.g. Alves et al. 2010). Alves et al. (2010, 2012) present maps of free-free using a RRL survey on the Galactic plane between longitudes $20° \geq l \geq 44°$. We will use these data in Chap. 3.

1.2.4 Free-Free Polarisation

Although the radiation from a single interaction is linearly polarised, the emission from a thermalised plasma will result in a zero net polarisation, as the collisions do not occur in any preferred direction. Nevertheless, Thompson scattering could yield a polarisation signal at the edges of HII regions, and its polarisation fraction can be as high as $\sim 10\%$ (Rybicki and Lightman 1979; Keating et al. 1998). Macellari et al. (2011) measure the free-free polarisation over the full sky, and found it to be less than 1%.

1.2.5 Anomalous Microwave Emission

In addition to the invaluable contribution to cosmology from the discovery of the CMB anisotropies, the analysis of the *Cosmic Background Explorer* (*COBE*) data brought surprises to the study of the local ISM in our Galaxy. Kogut et al. (1996a, b) statistically detected a correlated far-IR signal with a flat spectral index between 31 and 53 GHz. Then, observations at high Galactic latitudes by Leitch et al. (1997) showed emission at 14.5 GHz strongly correlated with *IRAS* 100 μm dust maps. This emission at $\nu > 10$ GHz was deemed anomalous because classical emission mechanisms could not account for it.

Since its discovery as a CMB-foreground, the dust-correlated anomalous microwave emission (AME) has been observed in different astrophysical environments, like molecular clouds (Finkbeiner et al. 2002; Watson et al. 2005; Casassus et al. 2006, 2008; AMI Consortium et al. 2009; Dickinson et al. 2010), translucent clouds (Vidal et al. 2011), HII regions (Dickinson et al. 2007, 2009; Todorović et al. 2010) and in the galaxy NGC 6946 by Murphy et al. (2010). Different emission mechanism have been proposed, such as spinning dust (Draine and Lazarian 1998; Ali-Haïmoud et al. 2009; Hoang et al. 2010; Ysard et al. 2011; Hoang and Lazarian 2012), magnetic dipole radiation (Draine and Lazarian 1999), hot ($T \sim 10^6$ K) free-free (Leitch et al. 1997) and flat spectrum synchrotron (Bennett et al. 2003). To date, the observations have favoured the spinning dust grain model (Finkbeiner et al. 2004; de Oliveira-Costa et al. 2004; Watson et al. 2005; Casassus et al. 2006, 2008). But it was only until recently that data from the *Planck* mission provided definitive evidence of spinning dust emission in the Perseus and ρ Oph molecular clouds (Planck Collaboration et al. 2011). The spectral energy distribution (SED) of these clouds between 10 and 100 GHz is well fitted only when a spinning dust component with physical properties compatible with the environment of the clouds is added. Figure 1.8 shows the spectral energy distribution (SED) of the Perseus cloud with the spinning dust component that fits the data. Planck Collaboration et al. (2013d) expands the analysis to a larger sample of AME regions, detecting 28 with high significance. They conclude that AME in these regions is also consistent with spinning dust emission.

Fig. 1.8 SED of the source
G160.2618.62 in the Perseus
molecular cloud. The *black
line* shows the best model to
the data which includes a
free-free (*orange-dash*),
thermal-dust (*cyan dashed
line*) and two-component
spinning dust model
(*magenta* and *green lines*).
Figure from Planck
Collaboration et al. (2013d)

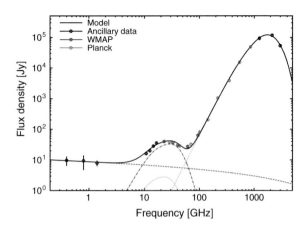

In the spinning dust scenario, very small grains (VSG) with a non-zero dipolar moment rotating at GHz frequencies emit radiation that peaks at ∼30 GHz. The models predict that the emission is dominated by the smallest grains ($\lesssim 0.001\,\mu$m), as they are more susceptible to the processes that affect the rotational dynamics of the grain. The main processes are gas-grain interactions, plasma-grain interactions, infrared emission and the electric dipole radio emission. The relative importance of these effects would depend in the environment of the grains. We will discuss the different aspects in the spinning dust modelling in Chap. 5.

1.2.6 AME Polarisation

A rotating dust grain would tend to align its angular momenta vector parallel to the interstellar magnetic field. Lazarian and Draine (2000) showed that this effect can be important for small grains ($\lesssim 0.05\,\mu$m). By this mechanism, the spinning dust emission would be polarised at a low level ($\lesssim 5\,\%$ at 10 GHz).

There is a small number of works in the literature that study the polarisation of AME. Battistelli et al. (2006) first measured the fractional polarisation in the Perseus AME dominated region to be $3.4^{+1.5}_{-1.9}\%$ at the 2σ level. Kogut et al. (2007) using the *WMAP* 3-year data limited the AME polarisation level to less than 1 %. Mason et al. (2009) gave a 2σ upper limit of 2.7 % at 9.65 GHz for LDN1622. Casassus et al. (2008) placed 3σ upper limits of 4.8 % on the polarisation of the ρ Oph cloud. López-Caraballo et al. (2011) using *WMAP* 7-yr data obtained 2σ upper limits of 1.0, 1.8 and 2.7 % at 23, 33 and 41 GHz respectively in the Perseus complex. Using the same *WMAP* data, Macellari et al. (2011) constrained the dust-correlated component of the polarised sky to less than 5 %. Dickinson et al. (2011) set upper limits of less than 2.6 % on the polarisation of ρ Oph and the Perseus region using *WMAP* data as well (see Sect. 2.6). More recently, Génova-Santos et al. (2011) and Rubiño-Martín et al. (2012) estimated upper limits in the Pleiades reflection nebulae and in the H II region

LPH96 and LDN1622 using *WMAP* data. Their 2σ constrains for the polarisation fraction at 23 GHz are 12.2, 1.3 and 2.6 % for each cloud respectively.

All these measurements imply a low level of polarisation of AME, below $\sim$3 %. This is consistent with expectations based on grain alignment by Lazarian and Draine (2000) and also rules out models for AME that predict high levels of polarisation, like the magnetic dipole emission from Draine and Lazarian (1999) where the polarisation fraction can be as high as 40 %.

1.2.7 Thermal Dust

Thermal emission from dust grains is the dominant emission mechanism in the Galaxy between $300 \lesssim \nu \lesssim 6000$ GHz ($50 \, \mu$m $< \lambda < 1$ mm). Dust grains are heated by starlight and this energy is re-radiated at longer wavelengths, mainly into the far-infrared. Large dust grains of sizes $\gtrsim 0.01 \, \mu$m are in equilibrium with the interstellar radiation field (IRF) and their temperature is in the range $10 \, \text{K} \lesssim T \lesssim 50 \, \text{K}$.

In a cloud in thermal equilibrium, the specific intensity from the dust grains at temperature T_d is given by:

$$I_\nu = B_\nu(T_d)[1 - e^{-\tau_\nu}] \tag{1.15}$$

where, $B_\nu(T_d)$ is the Planck function for a temperature T_d and τ_ν is the optical depth at a frequency ν. In the optically thin limit, i.e. $\tau_\nu << 1$, we have

$$I_\nu = B_\nu(T_d)\tau_\nu. \tag{1.16}$$

The optical depth τ_ν is the product between the dust emissivity cross section per mas unit $\kappa(\nu)$ (in units of cm^{-2} g^{-1}) and the column density of the dust mass,

$$\tau_d(\nu) = \kappa(\nu) \int \rho \, dl = \kappa(\nu) M_{\text{dust}}, \tag{1.17}$$

where ρ is the dust mass density along the line-of-sight and M_{dust} is the dust mass, that can be written in terms of the hydrogen column density, $N(H)$, as $M_{\text{dust}} = r \, \mu \, m_H \, N(H)$. r is the dust-to-gas mass ratio, μ is the mean molecular weight and m_H the mass of an Hydrogen atom. $\kappa(\nu)$ is usually described using a power law (e.g. Compiègne et al. 2011), $\kappa(\nu) = \kappa_0(\nu/\nu_0)^{\beta_d}$. Using these definitions, the dust specific intensity is written as,

$$I_\nu = B_\nu(T_d) \kappa_0 (\nu/\nu_0)^{\beta_d} r \, \mu \, m_H \, N(H). \tag{1.18}$$

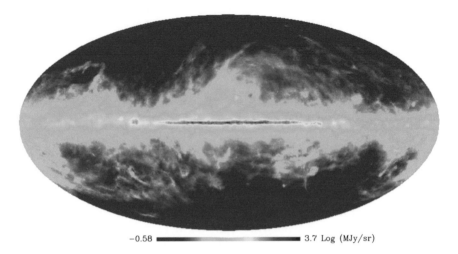

−0.58 ████████████████████████ 3.7 Log (MJy/sr)

Fig. 1.9 *Planck* 857 GHz map tracing the radiation from thermal dust. The *colour scale* is logarithmic to highlight the emission from the most diffuse areas at high Galactic latitudes

When modelling the dust emission from observations, it is common to use a simplified version of the previous equation, which has the form,

$$I_\nu = B_\nu(T_{obs})\, \tau_\nu\, (\nu/\nu_0)^{\beta_{obs}}, \tag{1.19}$$

where T_{obs} and β_{obs} are obtained from the data.

The *IRAS* satellite (Neugebauer et al. 1984) mapped the entire sky at four wavelengths bands centred at, 12, 25, 60 and 100 μm with an angular resolution better than 5 arcmin. The 100 μm map is a good tracer of the thermal emission from the large grains. The *Planck* spacecraft also provided six maps that are dominated by this emission from 100 to 857 GHz. In Fig. 1.9 we show the *Planck* 857 GHz map of the full sky, showing the distribution of the dust emission.

1.2.8 Thermal Dust Polarisation

Polarisation observations of starlight in the optical showed that dust grains are considerably non-spherical and also aligned with the interstellar magnetic field (Hiltner 1949; Hall and Mikesell 1949). The grains absorb the optical and UV starlight and then they re-radiate the energy in the far-infrared and submillimeter. If the grains are elongated and aligned, the total emission will present linear polarisation. In a simplified view, the grain alignment occurs in two steps. First, the grain axes align with respect to its angular moment and then, the angular moment aligns with the ambient magnetic field. The exact physics of alignment is complicated and different mechanisms are described in the review by Lazarian (2003).

The Archeops experiment measured for the first time the diffuse Galactic dust polarisation at 353 GHz over 17 % of the sky (Benoît et al. 2004). They found a polarisation fraction at the level of 4–5 %. Data from the BICEP experiment showed a fairly constant polarisation fraction $\sim$3 % between at 96, 15 and 210 GHz on two regions on the Galactic plane (Bierman et al. 2011). At higher frequencies, the polarisation fraction varies considerably and the slope of the spectra can even change sign, which probably means variations in the temperature of the dust populations along the line-of-sight (Hildebrand and Kirby 2004). Nevertheless, the polarisation angle reliably traces the magnetic field, because of the efficiency of the grain alignment processes with the interstellar field (Vaillancourt 2007; Lazarian and Finkbeiner 2003). This make dust polarisation a good tracer of the magnetic field in molecular clouds and more diffuse environments. Jaffe et al. (2013) used the polarised emission from dust in the *WMAP* 94 GHz band to test a large scale magnetic field model for the Galaxy. The more sensitive *Planck* polarisation data would help to constrain even further these models.

References

Ackermann, M., et al. (2012). Fermi-LAT observations of the diffuse γ-ray emission: implications for cosmic rays and the interstellar medium. *ApJ, 750*(3), 3.

Adriani, O., et al. (2011). Cosmic-ray electron flux measured by the PAMELA experiment between 1 and 625 GeV. *Physical Review Letters, 106*20, 201101.

Ali-Haïmoud, Y., Hirata, C. M., & Dickinson, C. (2009). A refined model for spinning dust radiation. *MNRAS, 395*, 1055–1078.

Alpher, R. A., & Herman, R. C. (1948). Evolution of the universe. *Nature, 162*, 774–775.

Alpher, R. A., & Herman, R. C. (1949). Remarks on the evolution of the expanding universe. *Physical Review, 75*, 1089–1095.

Alves, M. I. R. et al. (2010). Diffuse radio recombination line emission on the galactic plane between l = 36°. *MNRAS, 405*, 1654–1669.

Alves, M. I. R. et al. (2012). A derivation of the free-free emission on the galactic plane between l= 20°. *MNRAS, 422*, 2429–2443.

AMI Consortium et al. (2009). AMI observations of Lynds dark nebulae: further evidence for anomalous cm-wave emission. *MNRAS, 400*, 1394–1412.

Arnold, K. et al. (2010). The POLARBEAR CMB polarization experiment. In *Society of Photo-Optical Instrumentation Engineers (SPIE) Conference Series*. Vol. 7741.

Austermann, J. E. et al. (2012). SPTpol: an instrument for CMB polarization measurements with the south pole telescope. In *Society of Photo-Optical Instrumentation Engineers (SPIE) Conference Series*. Vol. 8452.

Banday, A. J., & Wolfendale, A. W. (1991). Fluctuations in the galactic synchrotron radiation. I - Implications for searches for fluctuations of cosmological origin. *MNRAS, 248*, 705–714.

Battistelli, E. S. et al. (2006). Polarization observations of the anomalous microwave emission in the perseus molecular complex with the COSMOSOMAS experiment. *ApJ, 645*, L141–L144.

Baumann, D. et al. (2009). Probing inflation with CMB polarization. In S. Dodelson et al. (Ed.), *American Institute of Physics Conference Series*. Vol. 1141, pp. 10–120.

Benford, D. J. et al. (2010). 5,120 superconducting bolometers for the PIPER balloonborne CMB polarization experiment. In *Society of Photo-Optical Instrumentation Engineers (SPIE) Conference Series*. Vol. 7741.

Bennett, C. L., et al. (1992). Preliminary separation of galactic and cosmic microwave emission for the COBE differential microwave radiometer. *ApJ*, *396*, L7–L12.

Bennett, C. L., et al. (2003). First-year Wilkinson microwave anisotropy probe (WMAP) observations: foreground emission. *ApJS*, *148*, 97–117.

Bennett, C. L., et al. (2013). Nine-year Wilkinson microwave anisotropy probe (WMAP) observations: final maps and results. *ApJS*, *208*(20), 20.

Benoît, A., et al. (2004). First detection of polarization of the submillimetre diffuse galactic dust emission by Archeops. *A & A*, *424*, 571–582.

Beringer, J. et al. (2012). Review of Particle Physics. *Physical Review D*, *86*(1), 010001.

BICEP1 Collaboration et al. (2013) Degree-scale CMB polarization measurements from three years of BICEP1 data. ArXiv e-prints.

Bierman, E. M., et al. (2011). A millimeter-wave galactic plane survey with the BICEP polarimeter. *ApJ*, *741*(81), 81.

Bischoff, C., et al. (2008). New measurements of fine-scale CMB polarization power spectra from CAPMAP at both 40 and 90 GHz. *ApJ*, *684*, 771–789.

Boggess, N. W., et al. (1992). The COBE mission - Its design and performance two years after launch. *ApJ*, *397*, 420–429.

Brown, M. L., et al. (2009). Improved measurements of the temperature and polarization of the cosmic microwave background from QUaD. *ApJ*, *705*, 978–999.

Carlstrom, J. E., et al. (2011). The 10 meter south pole telescope. *PASP*, *123*, 568–581.

Carretti, E. (2010). Galactic foregrounds and CMB polarization. In R. Kothes, T. L. Landecker, & A. G. Willis (Eds.), *Astronomical Society of the Pacific Conference Series*, Vol. 438, p. 276.

Casassus, S., et al. (2006). Morphological analysis of the centimeter-wave continuum in the dark cloud LDN 1622. *ApJ*, *639*, 951–964.

Casassus, S., et al. (2008). Centimetre-wave continuum radiation from the r Ophiuchi molecular cloud. *MNRAS*, *391*, 1075–1090.

Compiègne, M., et al. (2011). The global dust SED: tracing the nature and evolution of dust with DustEM. *A & A*, *525*(A103), A103.

Das, S. et al. (2013). The Atacama cosmology telescope: temperature and gravitational lensing power spectrum measurements from three seasons of data. ArXiv e-prints.

Davies, R. D., et al. (2006). A determination of the spectra of galactic components observed by the Wilkinson microwave anisotropy probe. *MNRAS*, *370*, 1125–1139.

de Oliveira-Costa, A., et al. (2004). The quest for microwave foreground X. *ApJ*, *606*, L89–L92.

Dicke, R. H., et al. (1965). Cosmic black-body radiation. *ApJ*, *142*, 414–419.

Dickinson, C., Davies, R. D., & Davis, R. J. (2003). Towards a free-free template for CMB foregrounds. *MNRAS*, *341*, 369–384.

Dickinson, C., Peel, M., & Vidal, M. (2011). New constraints on the polarization of anomalous microwave emission in nearby molecular clouds. *MNRAS*, *418*, L35–L39.

Dickinson, C., et al. (2007). CBI limits on 31 GHz excess emission in southern HII regions. *MNRAS*, *379*, 297–307.

Dickinson, C., et al. (2009). Anomalous microwave emission from the hii region RCW175. *ApJ*, *690*, 1585–1589.

Dickinson, C., et al. (2010). Infrared-correlated 31-GHz radio emission from orion east. *MNRAS*, *407*, 2223–2229.

Draine, B. T., & Lazarian, A. (1998). Electric dipole radiation from spinning dust grains. *ApJ*, *508*, 157–179.

Draine, B. T., & Lazarian, A. (1999). Magnetic dipole microwave emission from dust grains. *ApJ*, *512*, 740–754.

Dunkley, J., et al. (2009). Five-year Wilkinson microwave anisotropy probe (WMAP) observations: Bayesian estimation of cosmic microwave background polarization maps. *ApJ*, *701*, 1804–1813.

Eimer, J. R. et al. (2012). The cosmology large angular scale surveyor (CLASS): 40 GHz optical design. In *Society of Photo-Optical Instrumentation Engineers (SPIE) Conference Series*. Vol. 8452.

Essinger-Hileman, T. et al. (2010). The Atacama B-mode search: CMB polarimetry with transition-edge-sensor bolometers. ArXiv e-prints.

Finkbeiner, D. P. (2003). A full-sky Ha template for microwave foreground prediction. *ApJS, 146*, 407–415.

Finkbeiner, D. P., Langston, G. I., & Minter, A. H. (2004). Microwave interstellar medium emission in the green bank galactic plane survey: evidence for spinning dust. *ApJ, 617*, 350–359.

Finkbeiner, D. P., et al. (2002). Tentative detection of electric dipole emission from rapidly rotating dust grains. *ApJ, 566*, 898–904.

Fixsen, D. J. (2009). The temperature of the cosmic microwave background. *ApJ, 707*, 916–920.

Fixsen, D. J., & Mather, J. C. (2002). The spectral results of the far-infrared absolute spectrophotometer instrument on COBE. *ApJ, 581*, 817–822.

Fraisse, A. A., et al. (2013). SPIDER: probing the early universe with a suborbital polarimeter. *Journal of Cosmology and Astroparticle Physics, 4*(047), 47.

Frieman, J. A., Turner, M. S., & Huterer, D. (2008). Dark energy and the accelerating universe. *ARA & A, 46*, 385–432.

Génova-Santos, R., et al. (2011). Detection of anomalous microwave emission in the pleiades reflection nebula with Wilkinson microwave anisotropy probe and the cosmosomas experiment. *ApJ, 743*(67), 67.

Giardino, G., et al. (2002). Towards a model of full-sky galactic synchrotron intensity and linear polarisation: a re-analysis of the Parkes data. *A & A, 387*, 82–97.

Guzmán, A. E., et al. (2011). All-sky galactic radiation at 45 MHz and spectral index between 45 and 408 MHz. *A& A, 525*(A138), A138.

Hall, J. S., & Mikesell, A. H. (1949). Observations of polarized light from stars. *AJ, 54*, 187–188.

Hanson, D. et al. (2013). Detection of B-mode polarization in the cosmic microwave background with data from the south pole telescope. *Physical Review Letters, 111*(14), 141301.

Haslam, C. G. T. et al. (1982). A 408 MHz all-sky continuum survey. II - the atlas of contour maps. *A& AS, 47*, 1.

Hildebrand, R., & Kirby, L. (2004). Polarization of FIR/Sub-mm Dust Emission. In A. N. Witt, G. C. Clayton, & B. T. Draine (Eds.) Astrophysics of Dust, *Astronomical Society of the Pacific Conference Series*. Vol. 309, p. 515.

Hiltner, W. A. (1949). On the presence of polarization in the continuous radiation of stars. II. *ApJ, 109*, 471.

Hinshaw, G., et al. (2013). Nine-year Wilkinson microwave anisotropy probe (WMAP) observations: cosmological parameter results. *ApJS, 208*(19), 19.

Hoang, T., Draine, B. T., & Lazarian, A. (2010). Improving the model of emission from spinning dust: effects of grain Wobbling and transient spin-up. *ApJ, 715*, 1462–1485.

Hoang, T., & Lazarian, A. (2012). Acceleration of very small dust grains due to random charge fluctuations. *ApJ, 761*(96), 96.

Hu, W., & Dodelson, S. (2002). Cosmic microwave background anisotropies. *ARA & A, 40*, 171–216.

Hu, W., & White, M. (1997). A CMB polarization primer. *New Astronomy, 2*, 323–344.

Jaffe, T. R., et al. (2010). Modelling the Galactic magnetic field on the plane in two dimensions. *MNRAS, 401*, 1013–1028.

Jaffe, T. R., et al. (2013). Comparing polarized synchrotron and thermal dust emission in the galactic plane. *MNRAS, 431*, 683–694.

Keating, B. et al. (1998). Large angular scale polarization of the cosmic microwave background radiation and the feasibility of its detection. *ApJ, 495*, 580.

King, O. G. et al. (2013). The C-band all-sky survey (C-BASS): design and implementation of the northern receiver. ArXiv e-prints.

Kogut, A. et al. (1996a). High-latitude galactic emission in the cobe differential microwave radiometer 2 year sky maps. *ApJ, 460*, 1-+.

Kogut, A. et al. (1996b). Microwave emission at high galactic latitudes in the four- year DMR sky maps. *ApJ, 464*, L5+.

Kogut, A., et al. (2007). Three-year Wilkinson microwave anisotropy probe (WMAP) observations: foreground polarization. *ApJ, 665*, 355–362.

Larson, D., et al. (2011). Seven-year Wilkinson microwave anisotropy probe (WMAP) observations: power spectra and WMAP-derived parameters. *ApJS, 192*(16), 16.

Lawson, K. D. et al. (1987). Variations in the spectral index of the galactic radio continuum emission in the northern hemisphere. *MNRAS, 225*, 307.

Lazarian, A. (2003). Magnetic fields via polarimetry: progress of grain alignment theory. *Journal of Quantitative Spectroscopy & Radiative Transfer, 79*, 881.

Lazarian, A., & Draine, B. T. (2000). Resonance paramagnetic relaxation and alignment of small grains. *ApJ, 536*, L15–L18.

Lazarian, A., & Finkbeiner, D. (2003). Microwave emission from aligned dust. *New Astronomy Reviews, 47*, 1107–1116.

Leitch, E. M. et al. (1997). An anomalous component of galactic emission. *ApJ, 486*, L23+.

Leitch, E. M., et al. (2005). Degree angular scale interferometer 3 year cosmic microwave background polarization results. *ApJ, 624*, 10–20.

Liddle, A. R., & Lyth, D. H. (2000). Cosmological inflation and large-scale structure.

Linde, A. (2008). Inflationary cosmology. In M. Lemoine, J. Martin, P. Peter (Eds.), Inflationary Cosmology, *Lecture Notes in Physics*. Vol. 738, p. 1. Berlin: Springer.

Longair, M. S. (1994). High energy astrophysics. Vol. 2. Stars, the Galaxy and the interstellar medium.

López-Caraballo, C. H., et al. (2011). Constraints on the polarization of the anomalous microwave emission in the perseus molecular complex from seven-year WMAP data. *ApJ, 729*(25), 25.

Macellari, N., et al. (2011). Galactic foreground contributions to the 5-year Wilkinson microwave anisotropy probe maps. *MNRAS, 418*, 888–905.

Mason, B. S., et al. (2009). A limit on the polarized anomalous microwave emission of Lynds 1622. *ApJ, 697*, 1187–1193.

Miville-Deschênes, M.-A., et al. (2008). Separation of anomalous and synchrotron emissions using WMAP polarization data. *A& A, 490*, 1093–1102.

Montroy, T. E., et al. (2006). A measurement of the CMB ¡EE¿ spectrum from the 2003 flight of BOOMERANG. *ApJ, 647*, 813–822.

Murphy, E. J., et al. (2010). The detection of anomalous dust emission in the nearby galaxy NGC 6946. *ApJ, 709*, L108–L113.

Neugebauer, G., et al. (1984). The infrared astronomical satellite (IRAS) mission. *ApJ, 278*, L1–L6.

Niemack, M. D. et al. (2010). ACTPol: a polarization-sensitive receiver for the Atacama cosmology telescope. In *Society of Photo-Optical Instrumentation Engineers (SPIE) Conference Series*. Vol. 7741.

Noterdaeme, P. et al. (2011). The evolution of the cosmic microwave background temperature. Measurements of TCMB at high redshift from carbon monoxide excitation. *A& A, 526*, L7.

Ogburn IV, R. W. et al. (2010). The BICEP2 CMB polarization experiment. In *Society of Photo-Optical Instrumentation Engineers (SPIE) Conference Series*. Vol. 7741.

Oxley, P. et al. (2004). The EBEX experiment. In M. Strojnik (Ed.), Infrared Spaceborne Remote Sensing XII, *Society of Photo-Optical Instrumentation Engineers (SPIE) Conference Series*. Vol. 5543, pp. 320-331.

Penzias, A. A., & Wilson, R. W. (1965). A measurement of excess antenna temperature at 4080 Mc/s. *ApJ, 142*, 419–421.

Planck Collaboration et al. (2011). Planck early results. XX. New light on anomalous microwave emission from spinning dust grains. *A& A, 536*, A20.

Planck Collaboration et al. (2013a). Planck 2013 results. I. Overview of products and scientific results. ArXiv e-prints.

Planck Collaboration et al. (2013b). Planck 2013 results. XV. CMB power spectra and likelihood. ArXiv e-prints.

Planck Collaboration et al. (2013c). Planck 2013 results. XXII. Constraints on inflation. ArXiv e-prints.

Planck Collaboration et al. (2013d). Planck intermediate results. XV. A study of anomalous microwave emission in galactic clouds. ArXiv e-prints.

Platania, P., et al. (2003). Full sky study of diffuse galactic emission at decimeter wavelengths. *A& A*, *410*, 847–863.

QUIET Collaboration et al. (2011). First season QUIET observations: measurements of cosmic microwave background polarization power spectra at 43 GHz in the multipole range 25 ¡= l ¡= 475. *ApJ*, *741*(111), 111.

Reich, P., & Reich, W. (1988). A map of spectral indices of the galactic radio continuum emission between 408 MHz and 1420 MHz for the entire northern sky. *A& AS*, *74*, 7–20.

Rubiño-Martín, J. A. et al. (2012). Observations of the polarisation of the anomalous microwave emission: a review. *Advances in Astronomy*, *2012*, 351836.

Rybicki, G. B., & Lightman, A. P. (1979). Radiative processes in astrophysics.

Ryden, B. (2003). Introduction to cosmology.

Sachs, R. K., & Wolfe, A.M. (1967). Perturbations of a cosmological model and angular variations of the microwave background. *ApJ*, *147*, 73.

Scott, D., & Smoot, G. (2004). Cosmic background radiation mini-review. ArXiv Astrophysics e-prints.

Sheehy, C. D. et al. (2010). The Keck array: a pulse tube cooled CMB polarimeter. In *Society of Photo-Optical Instrumentation Engineers (SPIE) Conference Series*. Vol. 7741.

Sievers, J. L. et al. (2007). Implications of the cosmic background imager polarization data. *ApJ*, *660*, 976–987.

Smoot, G. F. et al. (1992). Structure in the COBE differential microwave radiometer first-year maps. *ApJ*, *396*, L1–L5.

Story, K. T. et al. (2012). A measurement of the cosmic microwave background damping tail from the 2500-square-degree SPT-SZ survey. ArXiv e-prints.

Strong, A. W., Moskalenko, I. V., & Ptuskin, V. S. (2007). Cosmic-ray propagation and interactions in the galaxy. *Annual Review of Nuclear and Particle Science*, *57*, 285–327.

Strong, A. W., Moskalenko, I. V., & Reimer, O. (2004). Diffuse galactic continuu m gamma rays: a model compatible with EGRET data and cosmic-ray measurements. *ApJ*, *613*, 962–976.

Sunyaev, R. A., & Zeldovich, Y. B. (1970). Small-scale fluctuations of relic radiation. *Ap& SS*, *7*, 3–19.

Swetz, D. S., et al. (2011). Overview of the Atacama cosmology telescope: receiver, instrumentation, and telescope systems. *ApJS*, *194*(41), 41.

Todorović, M. et al. (2010). A 33-GHz very small array survey of the Galactic plane from l = 27°. *MNRAS*, *406*, 1629–1643.

Vaillancourt, J. E. (2007). Polarized emission from interstellar dust. In M.-A. Miville-Desch.enes, & F. Boulanger (Eds.), *EAS Publications Series*. Vol. 23.

Vidal, M. et al. (2011). Dust-correlated cm wavelength continuum emission from translucent clouds z Oph and LDN 1780. *MNRAS*, *414*, 2424–2435.

Watson, R. A. et al. (2005). Detection of anomalous microwave emission in the Perseus molecular cloud with the COSMOSOMAS experiment. *ApJ*, *624*, L89–L92.

Wilson, T. L., Rohlfs, K., & Hüttemeister, S. (2009). *Tools of radio astronomy*. Springer.

Wu, J. H. P. et al. (2007). MAXIPOL: data analysis and results. *ApJ*, *665*, 55–66.

Ysard, N., Juvela, M., & Verstraete, L. (2011). Modelling the spinning dust emission from dense interstellar clouds. *A& A*, *535*(A89), A89.

Zaldarriaga, M., & Seljak, U. (1997). All-sky analysis of polarization in the microwave background. *Physical Review D*, *55*, 1830–1840.

Chapter 2
Analysis Techniques for *WMAP* Polarisation Data

In this chapter we describe the processing of the *WMAP* data and other full sky maps for the polarisation analysis. We start by describing the *WMAP* dataset. In Sect. 2.3 we describe the smoothed versions of the polarised maps and how we treated the noise after the smoothing. Next, in Sect. 2.5, we explain the polarisation bias correction scheme we used to obtain polarisation intensity maps. Section 2.5.5 shows the bias-corrected polarisation maps used later in this work. We then describe in Sect. 2.4 a filtering procedure that is used to remove the large scale emission from the maps in order to highlight the filamentary structure seen in the polarisation maps. Finally, in Sect. 2.6 we apply the de-biasing scheme for setting limits on the polarisation fraction of AME using *WMAP* data.

2.1 *WMAP* 9-Year Data

The description of the 9-year data can be found in Bennett et al. (2013). Here we summarise the main characteristics of the data. The 9-year data release encompasses observations from 00:00:00 UT 2001 August 10 to 00:00:00 UT 2010 August 10. The efficiency of the mission during this time is ∼98.4 %, where most of the data excluded belongs to intervals with low thermal stability.

The *WMAP* data processing scheme can be summarised as follows. The Time Ordered Data (TOD) are calibrated for each radiometer in order to convert the raw differential data into temperature units. This calibration is based on the CMB monopole temperature measured by *COBE* (Mather et al. 1999) and the dipole component, which is the Doppler-induced pattern product of the motion of the spacecraft in the CMB rest frame. A gain term (units of counts mK^{-1}) and baseline (units of counts) are used in this gain model. The 9-year analysis includes an additive ageing term for the radiometer in the form $m \Delta t + c$, where Δt is the elapsed mission time. The calibration scheme models the dipole variations with very small residuals. The estimate of the absolute calibration uncertainty takes a conservative value of 0.2 % (1σ) (Bennett et al. 2013). The efficiency factors of the two *WMAP* optical systems

© Springer International Publishing Switzerland 2016   25
M. Vidal Navarro, *Diffuse Radio Foregrounds*, Springer Theses,
DOI 10.1007/978-3-319-26263-5_2

differ slightly from one another. This accounts for time-independent *transmission imbalance factors* in the map making. There is good agreement between the 9-year values and the previous 7-year values within the uncertainties.

The mapping process of the calibrated TOD can be explained as follows. The TOD **d** for one particular radiometer can be described as

$$\mathbf{d} = \mathbf{M}\mathbf{t} + \mathbf{n}, \tag{2.1}$$

where **M** is the observing matrix, which contains information about the scan pattern and the beam shape. **M** converts the sky signal **t** into the TOD **d**. **n** is a vector representing the radiometer noise. The map is produced by inverting this equation and solving for **t**. The quality of the final map would depend on how well characterised are the constituents of the observation matrix.

The pointing of *WMAP* is computed using observations of Jupiter and Saturn, which are coupled to the orientation of the spacecraft using on-board star trackers. The estimated uncertainty on the pointing is $10''$.

The beam is characterised by azimuthally averaging beam maps obtained using observations of Jupiter. The 9-year data release includes a correction for the non-axisymmetric component of the beam for the Stokes I maps. It has been shown that the effect of asymmetric beams can affect the estimation of polarisation spectral indices (Wehus et al. 2013), so we will keep track of this effect later on. In Table 2.1 we list some characteristics of the *WMAP* instruments. The beam solid angle is listed for each channel as well as the FWHM of a Gaussian parametrisation of its shape. Also listed are the central frequency band and the bandwidth of each band. The central frequencies listed are for a point source with a temperature spectral index $\beta = -2.01$.

Table 2.1 *WMAP* characteristics

	K-Band	Ka-Band	Q-Band	V-Band	W-Band
Frequency (GHz)	22.69	32.94	40.62	60.52	92.99
Bandwidth (GHz)	5.5	7.0	8.3	14.0	20.5
Beam size[a] ($\times 10^{-5}$sr)	24.69	14.42	8.964	4.200	2.093
Beam width[b] (deg)	0.88	0.66	0.51	0.35	0.22
System temperature, T_{sys} (K)	29	39	59	92	145
Sensitivity (mK s$^{1/2}$)	0.8	0.8	1.0	1.2	1.6
$\sigma_0(I)$ (mK)[c]	1.429	1.466	2.188	3.131	6.544
$\sigma_0(Q, U)$ (mK)[c]	1.435	1.472	2.197	3.141	6.560

[a]Solid angle in azimuthally symmetrised beam profile
[b]FWHM of a Gaussian approximation to the beam profile
[c]Noise per observation in each pixel of the maps

2.1.1 **WMAP *9-Year Maps***

The *WMAP* data are available as full sky maps at the five frequency bands. The Stokes *I*, *Q* and *U* maps are the final compact representation of the 9-year data. These maps are used to characterise the foreground emission, create a map of the CMB anisotropy and to perform the cosmological analysis. They are provided in the HEALPix[1] pixelisation scheme (Górski et al. 2005) with $N_{side} = 512$, which corresponds to a pixel size of $\sim 7'$ (the sphere has $12 \times N_{side}^2$ pixels). The data are in units of thermodynamic temperature which relates to antenna (brightness) temperature in the following way,

$$T_A = \frac{x^2 e^x}{(e^x - 1)^2} T_{thermo} \tag{2.2}$$

where $x = (h\nu)/(kT_0)$ and $T_0 = 2.726\,\text{K}$ is the temperature of the CMB (Fixsen 2009).

Figure 2.1 shows a Mollweide projection of the full sky Stokes *I* maps at the angular resolution of each band listed in Table 2.1. The emission at K-band is dominated by diffuse Galactic foreground emission, mainly distributed along the Galactic plane. With increasing frequency (e.g. V and W bands), the level of foregrounds decreases, allowing the CMB to be clearly visible at high Galactic latitudes.

Figures 2.2 and 2.3 show the Stokes *Q* and *U* maps for each frequency. They are smoothed to a common resolution of 1°. In both *Q* and *U*, the strongest emission is observed at K-band. Figure 2.4 shows the polarisation amplitude, $P = \sqrt{Q^2 + U^2}$ for each frequency band. These maps are affected by polarisation bias and they have not been corrected for this effect. We will describe this bias in detail in Sect. 2.5.

Maps are supplied that describe the noise properties of the pixelated data, in terms of a number of observations (N_{obs}) parameter. The total intensity noise maps for each frequency can be obtained using the following relation,

$$\sigma_I^2 = \frac{\sigma_0^2(I_p)}{N_{obs}(I_p)} \tag{2.3}$$

with $\sigma_0(I_p)$ the noise per observation value in each pixel, listed in Table 2.1.

The noise in the polarisation maps is characterised by the covariance matrix, which includes the correlation term between *Q* and *U*. This correlation occurs mainly due to non-uniform azimuthal coverage for each pixel and fluctuations in the noise during the observations. The correlation is position-dependent and is characterised by the $N_{obs}(QU)$ map. The covariance matrix for each pixel has the form,

$$\mathbf{C} = \begin{pmatrix} \sigma_Q^2 & \sigma_{QU} \\ \sigma_{QU} & \sigma_U^2 \end{pmatrix} = \sigma_0^2(QU) \begin{pmatrix} 1/N_{obs}(QQ) & 1/N_{obs}(QU) \\ 1/N_{obs}(QU) & 1/N_{obs}(UU) \end{pmatrix} \tag{2.4}$$

The $\sigma_0^2(QU)$ values for each band are also listed in Table 2.1.

---

[1]http://healpix.sourceforge.net/.

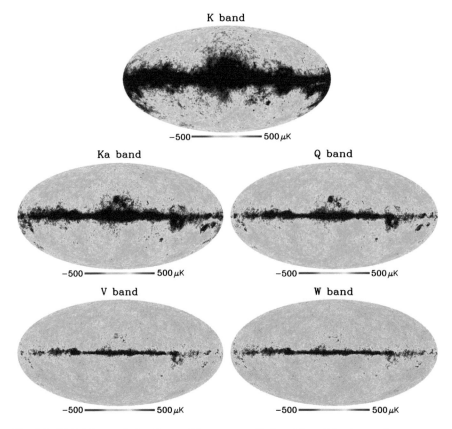

Fig. 2.1 *WMAP* 9-year Stokes *I* maps. The maps are displayed in a Galactic coordinate system using a Mollweide projection. The Galactic plane runs horizontally along the centre of each map. The Galactic longitude increases towards the left. The linear scale is the same in all maps, ranging from −500 to 500 μK in thermodynamics units. These maps have the original angular resolution at each band, as listed in Table 2.1

In Fig. 2.5, we show the noise maps for K-band, where the σ_I, σ_Q, σ_U and σ_{QU}^2 maps are shown. The structure in these maps is revealing the spacecraft observing pattern during the 9 year of data collection. The large circular features visible in them are centred at the ecliptic poles, which are the regions of the sky with the largest integration time. Similar maps exist for all the frequency bands but due to their similarity, here we only show the ones corresponding to K-band.

We note that the noise maps shown here do not include any systematic effects. Fluctuations in the gain of the receivers, thermal instabilities and noise temperature fluctuations in the TOD will produce correlated noise in the final maps. This type of noise, referred as "1/f noise" is not normally distributed. In general, the instrumental noise of *WMAP* data (or any similar instrument) can be described as the sum of white (Gaussian) and correlated noise. The power spectral density of the noise has the form,

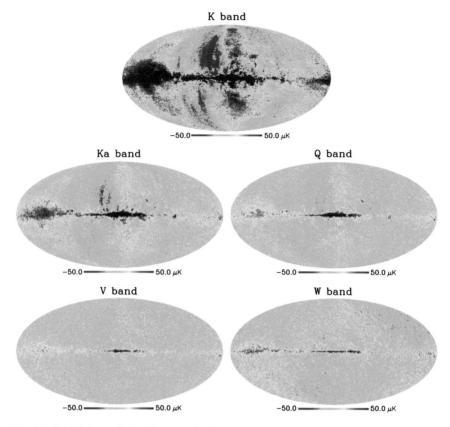

Fig. 2.2 *WMAP* 9-year Stokes Q maps. The linear scale is the same in all maps, ranging from -50 to $50\,\mu$K in thermodynamics units. The maps have been smoothed to a common resolution of $1°$

$$P_{\text{noise}} = \left[1 + \left(\frac{f_k}{f}\right)^{\alpha}\right]\frac{\sigma}{f_s}, \tag{2.5}$$

where f_k is called the "knee" frequency, f_s is the sampling frequency of the data and σ is the standard deviation of the white noise during the sample integration time ($t = 1/f_s$). For *WMAP* data, the $1/f$ component of the noise is modelled and subtracted from the data, leaving small residuals. The covariance along the scan directions is limited to be less than 0.1 % for the K, Ka, Q and V bands (Hinshaw et al. 2003). The W–band presents the worst $1/f$ noise, which is at least a factor 3 larger than the rest of the bands (Jarosik et al. 2003). This small contribution from "1/f" noise allows us to use only the white noise maps shown here to quantify the uncertainties of the data.

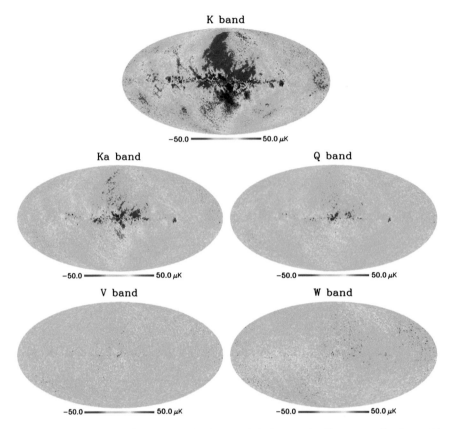

Fig. 2.3 *WMAP* 9-year Stokes *U* maps. The linear scale is the same in all maps, ranging from −50 to 50 μK in thermodynamics units. The maps have been smoothed to a common resolution of 1°

2.2 *WMAP* Sky Simulation

In order to test the results that we will present in this chapter and also in Chap. 3, we used a simulation of the synchrotron emission in *WMAP* data using the *Planck* Sky Model.

The Planck Sky Model (PSM) (Delabrouille et al. 2013), is a software package that implements modelling of the emission mechanisms at GHz frequencies with the aim of creating a realistic representation of the sky emission in both total intensity and polarisation. The PSM can model the CMB emission, SZ effect, extragalactic sources and the Galactic emission. There are five diffuse Galactic components: synchrotron, free-free, AME, thermal dust and CO molecular lines. In this case, we have simulated only the synchrotron emission as it is dominant in *WMAP* polarisation data. The synchrotron simulation is based on the model by Miville-Deschênes et al. (2008), which uses *WMAP* seven-year data and the 408 GHz map by Haslam et al. (1982) to fix its parameters. The simulations are created using a resolution parameter $N_{\text{side}} = 512$.

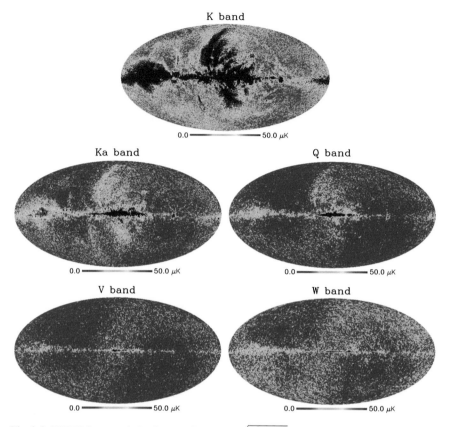

Fig. 2.4 *WMAP* 9-year polarisation amplitude $P = \sqrt{Q^2 + U^2}$ maps. The linear scale is the same in all maps, ranging from 50 to 50 μK in thermodynamics units. The maps have been smoothed to a common resolution of 1°

Figure 2.6 shows the templates for Stokes I, Q, U and the polarisation amplitude at K-band as observed by *WMAP* . There is no noise added to these simulations. The Q, U and P maps shown here are similar to the *WMAP* polarisation data shown in Figs. 2.2, 2.3 and 2.4.

2.3 Smoothing

To increase the signal-to-noise-ratio (SNR) of the maps, and to minimise any systematic effect due to beam-asymmetries in polarisation, we smoothed the maps to common resolutions of 1° and 3° full-width-half-maximum (FWHM). This was done using HEALPIX by first deconvolving in harmonic space the azimuthally symmetrized effective beam and then convolving with a Gaussian beam. The *WMAP* team

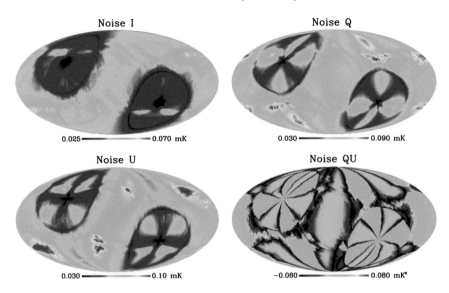

Fig. 2.5 *WMAP* 9-year K-band noise per pixel maps. On *top*, are the σ_I and σ_Q maps, and at the *bottom*, the σ_U and σ_{QU}^2 maps. The structure in these maps is related to the *WMAP* scanning pattern

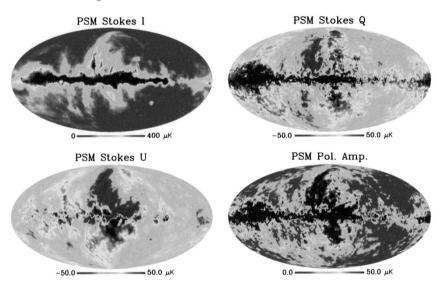

Fig. 2.6 Planck Sky Model templates for synchrotron emission as observed by *WMAP* at K-band. On *top* are Stokes *I* and *Q*, at the *bottom* are Stokes *U* and the polarisation amplitude map. These simulations do not include noise

provides the symmetrized beam transfer functions derived from the observations of Jupiter, B^S_ℓ, which describe the response of the antenna as a function of multipole. For the precision that the cosmological analysis requires, the *WMAP* beams have been very well characterised. We take advantage of the available transfer functions, instead of the commonly used Gaussian approximation for the beam. The convolution of the maps with a Gaussian beam corresponds to a multiplication in harmonic space of the $a_{\ell,m}$ representation of the map with an "effective" transfer function, defined by,

$$B^{eff}_\ell = e^{-\frac{1}{2}\ell(\ell+1)\sigma^2}/B^S_\ell, \tag{2.6}$$

where σ is the width of the smoothing Gaussian beam in radians.[2]

In Fig. 2.7 we show the symmetrised transfer functions for each *WMAP* band. Note the difference between the symmetrised beam transfer function in black, and the Gaussian parametrisation of the beam based on a Gaussian approximation in red, with a width equal to the values listed in Table 2.1. The largest deviations from a Gaussian beam occur for the higher frequency bands. The blue line is the effective transfer function used for the smoothing to obtain a map with an angular resolution of 1° FWHM.

For the spectral index analysis that we will present in the next chapter, we smoothed the polarisation maps to a common angular resolution of 3°. The choice of this smoothing scale is a compromise between having a good SNR across a large area of the sky and having enough spatial resolution to characterise the observed structures. Wehus et al. (2013) showed that the beam asymmetries in *WMAP* K- and Ka-bands give rise to unstable spectral index measurements in polarisation on 1° scales. Here, with an effective resolution of 3°, the asymmetries have minimal impact.

The propagation of the uncertainty in the maps after smoothing with a Gaussian kernel $G(\sigma^2)$ defined by its standard deviation σ, can be calculated analytically to create a new "smoothed" variance map N' if the errors are independent and normally distributed,

$$N'^2 = \frac{\Omega}{4\pi\sigma^2}N^2 \otimes G(\sigma^2/2), \tag{2.7}$$

with Ω the solid angle of each pixel. Note that the smoothing width, $\sigma/\sqrt{2}$, is different from the smoothing kernel of the map (σ).

Unfortunately, this is not valid for *WMAP* data. The noise is correlated at large angular scales. Moreover, in polarisation the covariance matrix is not diagonal as there are correlations between the uncertainties in Q and U. In order to asses the noise level after the smoothing of the maps, we used Monte Carlo simulations, which are described in the next Section.

---

[2]Note that σ relates with the FWHM by the relation $\sigma = \mathrm{FWHM}/\sqrt{8\ln 2}$.

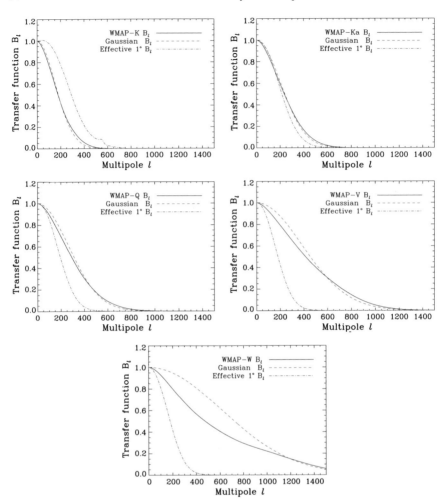

Fig. 2.7 The plots show the beam transfer functions as function of multipole for the five *WMAP* channels. In *black* is the symmetrised beam function as characterised by the *WMAP* team. In *red* is the Gaussian transfer function with a width equivalent to the ones tabulated in Table 2.1. Note the discrepancies between the "true" beam shape in *black* and the equivalent Gaussian in *red*. In *blue* is the effective beam transfer that we used to smooth the maps to a common 1° angular resolution FWHM

2.3.1 WMAP *Noise Simulations*

We ran Monte Carlo simulations to calculate the uncertainties after the smoothing and pixel downgrading. We generated 500 normally distributed noise realisation for Stokes *I*, *Q* and *U* for each frequency band using the covariance matrix from the

WMAP data. The covariance matrix is constructed for each pixel in the following way:

$$\mathbf{C} = \sigma_0^2(QU) \begin{pmatrix} N_{QQ} & N_{QU} \\ N_{QU} & N_{UU} \end{pmatrix}^{-1} \tag{2.8}$$

where $\sigma_0^2(QU)$ is the nominal polarisation noise per observation listed in Table 2.1 for each frequency band and N_{QQ}, N_{QU} and N_{UU} are the N_{obs} maps supplied with the data. We used the IDL routine MRANDOMN to draw each realisation of the noise. This routine uses a Cholesky decomposition of the covariance matrix to efficiently generate random deviates from a multivariate normal distribution. The output of the simulation are 500 noise maps for Stokes I, Q and U.

We calculated the statistics of this assembly and we used the dispersion on each pixel (for the I, Q, and U maps) as the statistical error on the 3° smoothed maps. We also calculated the covariance term between Q and U for the ensemble (σ_{QU}^2). By this route, we have smoothed maps for Stokes I, Q and U with their respective uncertainties and (Q, U) correlations for each pixel. For all the noise values estimated, we need to add in quadrature the 0.2 % *WMAP* absolute calibration error (Bennett et al. 2013).

2.3.2 Optimal N_{side}

After smoothing, the 3° resolution maps at $N_{side} = 512$ are over-sampled, i.e. there are too many pixels ($\sim$778) per beam area. Ideally, the pixel size should be similar to the beam size. The re-pixelisation onto a coarser grid (less and bigger pixels) of a scalar quantity, such as the total intensity brightness temperature is straightforward, the values of the smaller pixels are averaged into a single value that is given to the new, larger pixel.

When working with polarisation, more care is required. The polarisation of any given pixel has an amplitude and a direction, and both quantities should be adequately transformed onto the coarser grid. The standard re-pixelisation procedure (averaging of smaller pixels to form a new larger one) can induce systematic errors in the new polarisation maps, which are more prominent close to the coordinates poles. Parallel transport has to be used in order to produce adequate re-pixelisation of polarisation maps. The HEALPIX package includes a routine (`ud_grade`) to upgrade/degrade a full sky map in the HEALPIX scheme. Unfortunately, this routine does not include parallel transport and therefore, is not reliable close to the poles.

An alternative to downgrade the maps in this case is to generate new maps, in a new grid, from the pseudo-scalar E and B-modes spectra. To do this, we first used the HEALPIX routine `anafast` to create the power spectra of the polarisation maps. We save the $a_{\ell,m}$ coefficients and they are fed to the routine `synfast`, which creates a new map using the power spectra in a grid defined by the user. By this way, the new map is created with correct polarisation information. A problem with this method is that the fast Fourier transform (FFT) involved in the computation of the power spectra produces artefacts in the final map, which are more important around sharp features of the map (e.g. strong point sources, the Galactic plane).

In practise, the effect of using or not parallel transport is very small and it is only important close to the poles. In Fig. 2.8 we compare two polarisation maps P_1 and P_2 produced with the two methods described above. Both maps are downgraded versions of a 3° smoothed map from an original $N_{side} = 512$ to $N_{side} = 32$. P_1 is produced with the ud_grade routine without parallel transport; P_2 is produced with anafast + synfast, regridding the $a_{\ell,m}$ onto a coarser grid. The figure shows the difference between P_1 and P_2 divided by the polarisation noise σ_P, calculated simply as $\sigma_P = \sqrt{Q^2\sigma_Q^2 + U^2\sigma_U^2 + 2QU\sigma_{QU}^2}/P$ (this is not an accurate quantification of the polarisation noise as the quantity $P = \sqrt{Q^2 + U^2}$ has a positive bias which will be discussed in detail in Sect. 2.5, nevertheless, is a good approximation for the display purposes here).

In Fig. 2.8, the most obvious differences occur at the regions with larger signal, like the Galactic plane. They are due to FFT artefacts in the map produced using anafast + synfast. The level of these artefacts will be a function of the ℓ_{max} values adopted in anafast. We have used $\ell_{max} = 3 \cdot N_{side} - 1$. This way, the spherical harmonics form a linearly independent system. The difference between the two studied maps is small, the scale in the figure shows the $\pm 0.5\%$ range in terms of the polarisation noise. From the lower panel of the figure, we see the red blobs at the Galactic poles. This difference between the two maps at the poles is due to the parallel transport difference between the two polarisation maps. As this effect is smaller than the larger artefacts seen on the Galactic plane, we can safely ignore them in our analysis.

For the rest of this work, we will use the maps degraded using ud_grade, ignoring the parallel transport issues, as they are relatively small compared with the artefacts produced by the second method.

Test on spectral index estimation with different pixel sizes

We tested the effect of different pixel sizes on the estimation of the polarised spectral index (in Sect. 3.3 we discuss the polarised spectral indices in detail). For this, we used the PSM simulation at K-band described in Sect. 2.2. We scale this polarisation template to 33 GHz using an input spectral index $\beta = -3.0$. Then, we added a noise realisation using the *WMAP* covariance matrix. The idea of this test is to check if the measured spectral index and its uncertainty from these simulated maps varies with different pixel sizes.

The spectral index is measured over the entire map using the T-T plots approach (see Sect. 3.3 for a detailed description on this). The full-sky mean value and its uncertainty as a function of the pixel size is plotted in Fig. 2.9. In this figure, the measured spectral indices and their uncertainty are not visibly affected by the pixel size used. Here, the measured central value for β is not exactly the input value of $\beta = -3.0$, this is because used only one noise realisation. Nevertheless, what it is important is the consistency of the measured values with different pixel size.

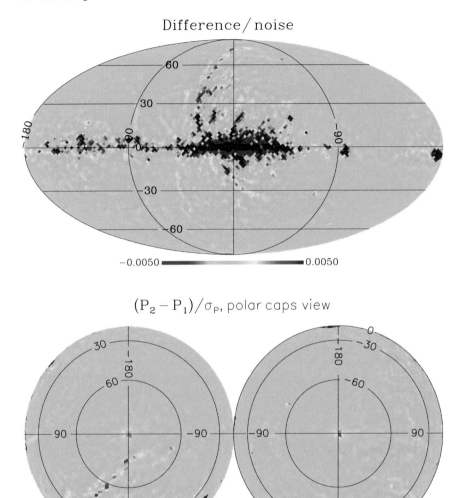

Fig. 2.8 Comparison between two *WMAP* K-band polarisation maps degraded from $N_{side} = 512$ to $N_{side} = 32$ using two methods: *i.* the routine ud_grade, which does not take into account parallel transport and *ii.* a re-mapping from the power spectra using the anafast and synfast routines. The difference is then divided by the uncertainty in the polarisation amplitude, σ_P. On *top* is a Mollweide projection showing the entire sky map. At the *bottom* is an orthographic projection centred at the north and south Galactic poles. Note the differences around the poles, seen as a two *red circular* regions

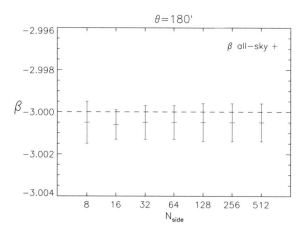

Fig. 2.9 Variation of the uncertainty in the polarisation spectral index as a function of the resolution parameter N_{side} for 3° resolution maps. β is calculated using 2 simulated maps at K– and Ka-band with a fiducial spectral index $\beta = -3.0$. The quoted error bar shows the error on the mean of the measured spectral index on the entire sky. There is no variation on the measured spectral index with different pixel size

This analysis allows us to choose the optimal pixel size that keeps the information of the spatial structures and allow us faster computation and analysis of the maps. This is $N_{side} = 32$ for the maps smoothed to 3°. For the maps that are smoothed to 1°, we use $N_{side} = 64$, which corresponds to pixel size $\approx 54'.9$.

2.4 Unsharp Mask

A convenient image-processing technique can be used in order to highlight certain features from a sky map. Here we describe the *unsharp masking* technique, which works as a spatial high-pass filter, washing-out features that are larger than a defined angular scale. We will use this technique in Chap. 3 to identify the filaments visible in the *WMAP* polarisation maps.

We start with an initial map of brightness temperature T. This map T is convolved with a Gaussian beam $G(\sigma)$ to produce $\widetilde{T}_0$:

$$T \otimes G = \widetilde{T}_0. \tag{2.9}$$

We then define:

$$\delta T_0 = T - \widetilde{T}_0, \tag{2.10}$$

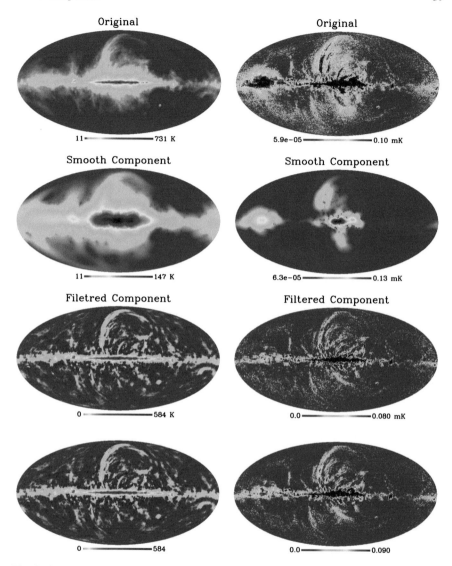

Fig. 2.10 Images illustrating the unsharp mask technique on two full-sky maps, the 408 MHz Haslam et al. (1982) map on the *left* column and the *WMAP* K-band polarisation map on the *right*. On *top* are shown the original maps, both of them smoothed to a common resolution of 1°. In the *middle*, is the smooth component, calculated using the iterative process described in the text. At the *bottom* are the filtered maps, highlighting the filamentary structure present in both maps. The convolution kernel has a size of 5° FWHM

which maps the small-scale (compared to the smoothing kernel) component of the original map. We also define

$$T_1 = \begin{cases} T - \delta T_0, & \text{if} \quad \delta T_0 > 0 \\ T, & \text{if} \quad \delta T_0 \leq 0. \end{cases} \tag{2.11}$$

T_1 is equal to the original map T except in the small-scale structures ($\delta T_0 > 0$), where it takes the value of $\widetilde{T}_0$.

T_1 is then convolved with the same Gaussian kernel $G(\sigma)$ to produce $\widetilde{T}_1$, ΔT_1 and T_2. This process is repeated i-times, until T_i converges to a unique background or until,

$$|T_i - T_{i-1}| < \sigma_T, \tag{2.12}$$

where σ_T is the RMS noise per pixel of the map, based in the number of observations in *WMAP* data. At this stage, T_i will be a template of the smooth large scale component of the original map. The unsharp-masked map is then defined by $T_{unsharp} = T - T_i$.

In Fig. 2.10 we show two examples of the process applied to the 408 MHz map from Haslam et al. (1982) map and to the *WMAP* K-band polarisation amplitude map. Both maps are initially smoothed to a common resolution of 1°. Then, the unsharp masking processes is executed, using a 5° Gaussian as the smoothing kernel. In the figure, the original maps (T), the large scale component (T_i) and the final filtered maps ($T_{unsharp}$) are shown. We will discuss and use these filtered maps in the next chapter.

2.5 Polarisation Bias

It has long been noticed that observations of linear polarisation are subject to bias (Serkowski 1958). Given the positive nature of $P = \sqrt{Q^2 + U^2}$, even if the true Stokes parameters are zero, P will yield a non-zero estimate in the presence of noise. The effect is particularly important in the low SNR regime. In this section we review different estimators used in the literature, propose an alternative one, more suitable for *WMAP* data and finally we test the efficiency of the different methods using Monte Carlo simulations.

Ways to correct for the bias have been studied in detail (Quinn 2012; Simmons and Stewart 1985; Vaillancourt 2006; Wardle and Kronberg 1974) for the special case where the uncertainties for (Q, U) are equal and normally distributed around their true value (Q_0, U_0). We will first review this case and then study the more general case with unequal uncertainties in (Q, U).

2.5.1 Symmetric Uncertainties

Let us take (Q_0, U_0) as the true Stokes parameters from a source and (Q, U) the measured ones. We can write the joint probability distribution function (p.d.f.) for (Q, U) as the product of the individual normal distributions

$$f(Q, U) = \frac{1}{2\pi\sigma^2} \exp\left(-\frac{(Q - Q_0)^2 + (U - U_0)^2}{2\sigma^2}\right), \qquad (2.13)$$

where $\sigma_Q = \sigma_U = \sigma$ is the uncertainty in (Q, U).

Transforming into polar coordinates using the definitions from Eq. 1.5,

$$
\begin{aligned}
f(P, \chi) &= \frac{P}{\pi\sigma^2} \exp\left(-\frac{P^2 + P_0^2 - 2(P\cos 2\chi\, P_0 \cos 2\chi_0 + P\sin 2\chi\, P_0 \sin 2\chi_0)}{2\sigma^2}\right) \\
&= \frac{P}{\pi\sigma^2} \exp\left(-\frac{P^2 + P_0^2}{2\sigma^2}\right) \exp\left(-\frac{PP_0 \cos[2(\chi - \chi_0)]}{\sigma^2}\right).
\end{aligned}
\qquad (2.14)
$$

The marginal probability distribution for P is obtained by integrating $f(P, \chi)$ over χ. This angular integral can be written as a function of the modified Bessel function of first type $I_0(z)$ (Vinokur 1965), yielding the Rice distribution for polarisation

$$R(P|P_0) = \frac{P}{\sigma^2} I_0\left(\frac{PP_0}{\sigma^2}\right) e^{-\frac{P^2 + P_0^2}{2\sigma^2}}. \qquad (2.15)$$

It is important to note that the integral of $R(P|P_0)$ represents the probability of measuring P inside an interval for a *known* true polarisation P_0. Figure 2.11 shows $R(P|P_0)$ for different SNR. The bias, defined as $\langle P \rangle - P_0$, arises as this probability distribution is not symmetric and becomes clear in Fig. 2.11 at low SNR levels. Even

Fig. 2.11 Rice distribution (Eq. 2.15) plotted for different values of the true SNR, P_0/σ. The asymmetry and bias are clear in the low SNR level. At high SNR, the distribution converges to a Gaussian with standard deviation σ centred at P_0

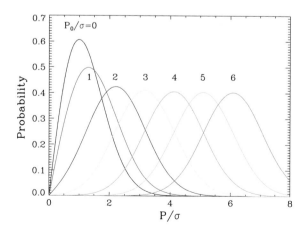

when the true SNR is zero (black curve in the Figure), the measured value is close to 1. At large SNR, the mean of the distribution approaches a Gaussian with mean close to P_0 and standard deviation close to σ.

If Eq. 2.14 is integrated with respect to P, it yields the probability distribution of the polarisation angle χ. This distribution for χ is given by Naghizadeh-Khouei and Clarke (1993) and Quinn (2012):

$$G(\chi|\chi_0, P_0, \sigma) = \frac{1}{\sqrt{\pi}} \left(\frac{1}{\sqrt{\pi}} + \eta_0 e^{\eta_0^2} [1 + \mathrm{erf}(\eta_0)] \right) e^{-\frac{P_0^2}{2\sigma^2}}, \tag{2.16}$$

with $\eta_0 = \frac{P_0}{\sqrt{2}\sigma} \cos[2(\chi - \chi_0)]$ and erf is the Gaussian error function.

Simmons and Stewart (1985) describe four estimators of P_0 that are better surrogates for P.

- The median estimator $\hat{p}_M$ defined as the value for which the observed value P equals the median of the distribution $R(P|P_0)$.
- The mean estimator $\hat{p}_S$ defined as the value for which the observed value P equals the mean of the distribution $R(P|P_0)$.
- The Wardle and Kroberg's estimator $\hat{p}_{WK}$ which is obtained when the observed polarisation is a maximum of $R(P|P_0)$, that is

$$\left. \frac{\partial R}{\partial P}(P, P_0) \right|_{P_0 = \hat{p}_{WK}} = 0. \tag{2.17}$$

Therefore, $\hat{p}_{WK}$ satisfies

$$I_0 \left(\frac{\hat{p}_{WK}P}{\sigma^2} \right) \left(1 - \frac{\hat{p}_{WK}^2}{\sigma^2} \right) + \frac{\hat{p}_{WK}P}{\sigma^2} I_1 \left(\frac{\hat{p}_{WK}P}{\sigma^2} \right) = 0. \tag{2.18}$$

Wardle and Kronberg (1974) propose an analytic approximation to solve this equation valid for $P/\sigma > 1$. The estimator

$$\hat{p}_{WK} \approx \sqrt{P^2 - \sigma^2} \tag{2.19}$$

performs accurately for SNR $\gtrsim 2$.

- The maximum likelihood estimator $\hat{p}_{ML}$, is the value of P_0 that maximises $R(P|P_0)$ for a given observed polarisation P, that is

$$\left. \frac{\partial R}{\partial P_0}(P, P_0) \right|_{P_0 = P_{WK}} = 0. \tag{2.20}$$

Note that this is not equivalent to Eq. 2.17 as the differentiation in this case is with respect to P_0. $\hat{p}_{ML}$ can be found by evaluating numerically the following expression for a given P,

$$PI_0 \left(\frac{\hat{p}_{ML}P}{\sigma^2} \right) - \hat{p}_{ML}I_1 \left(\frac{\hat{p}_{ML}P}{\sigma^2} \right) = 0. \tag{2.21}$$

Simmons and Stewart (1985) conclude that the maximum likelihood estimator is superior for low SNR ($P_0/\sigma \lesssim 0.7$) and the Wardle and Kroberg's estimator $\hat{p}_{WK}$ performs better for higher SNR. The median and mean estimator are the best only in a very narrow SNR region. When the SNR is large, ($P_0/\sigma > 4$), all estimators converge. In this case, it is more convenient the simpler representation of $\hat{p}_{WK}$ (Eq. 2.19).

2.5.2 Asymmetric Uncertainties

As we saw, the previous case in which the uncertainties in (Q, U) are equal is well understood and solved for a fair range of SNR. The asymmetric case ($\sigma_Q \neq \sigma_U$) is interesting as many polarisation data sets have this characteristic. In particular, the correlations between the (Q, U) uncertainties in *WMAP* data are mainly due to non-uniform azimuthal coverage for each pixel in the sky and fluctuations in the noise during the observations. The error probability distribution in this case is an elliptical 2D Gaussian in (Q, U), in which the semi-axis correspond to the uncertainties in (Q, U), σ_Q and σ_U. In Fig. 2.12 are shown the histograms of the eccentricity[3] of the error ellipses for all the pixels in the five *WMAP* bands.

When the uncertainties in (Q, U) are not equal, the polarisation bias depends also in the polarisation angle χ as well as on p. A generalisation of the Wardle and Kronberg estimator can be written for this case. It has the form:

$$\hat{p}_{wk} = \sqrt{P'^2 - (\sigma_Q^2 \sin^2 2\chi' + \sigma_U^2 \cos^2 2\chi' - 2\sigma_{QU} \cos 2\chi' \sin 2\chi')}. \tag{2.22}$$

This estimator reduces to Eq. 2.19 when the errors are isotropic. Here, the observed polarisation angle, χ' is used as a surrogate for the true polarisation angle χ. At the high SNR regime (e.g. the bright regions in *WMAP* K-band), the approximation $\chi' \approx \chi$ is excellent. In the low SNR case, the answer is not straightforward and the only way to test this is to measure the residual bias of the estimator. We will discuss simulations that we use to measure the performance of this estimator in Sect. 2.5.4.

The estimator in Eq. 2.22 can be used to de-bias the polarisation maps at K-band, where there is a reasonable SNR along large areas of the sky. For the rest of the bands, where the SNR is much lower we can derive a new estimator. For this, we use the fact that we can obtain a good estimation of the *true* polarisation angle χ from *WMAP* K-band. Moreover, at *WMAP* frequencies, Faraday rotation is negligible over most of the sky (see Sect. 3.4) so we can assume that the polarisation angle χ

---

[3]The eccentricity, e, is defined as $e = \sqrt{1 - \left(\frac{\sigma_{min}}{\sigma_{maj}} \right)^2}$, where $\sigma_{min}/\sigma_{maj}$ correspond to the minor/major axial ratio of the ellipse.

Fig. 2.12 Histograms of the eccentricity of the polarisation probability distribution of *WMAP* data. The histograms are made using the full sky maps at the original angular resolution (see Table 2.1), with $N_{side} =$ 512

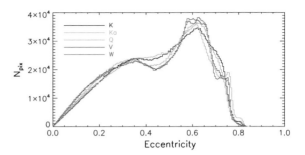

observed at K-band, will be the same at the higher frequency bands. The additional information about the polarisation angle in the higher frequency bands will help to reduce the total bias in the pixels with low SNR. In the next section we derive this new estimator, assuming that we know the true polarisation angle χ. In Sect. 2.5.4 we use simulations to test the effectiveness of this new estimator and to quantify any residual bias due to noise in χ.

2.5.3   Known Angle Estimator

We will build a new estimator assuming that the true polarisation angle χ is know, without any bias or noise. Later we will test the effect of these assumptions with simulations. This estimator can be used when a good independent measurement of the polarisation angle is available.

From Bayes theorem, the posterior p.d.f. for P, given the observed values (P', χ') is given by

$$f(P|\mathbf{P}', \chi') = \frac{f(\mathbf{P}', \chi'|P)f(P', \chi')}{\int f(\mathbf{P}', \chi'|P)f(P', \chi')dP'd\chi'} \qquad (2.23)$$

where $f(\mathbf{P}', \chi'|P) \equiv L(P)$ is the likelihood function of P and $f(P', \chi')$ is the prior p.d.f. for P.

Knowing χ and assuming a uniform prior for P (all values for the amplitude of P are equally probable), we have

$$f(P|\mathbf{P}', \chi_0) \propto f(\mathbf{P}', \chi_0|P). \qquad (2.24)$$

The maximum likelihood estimator in this case, $\hat{p}_{\chi_0}$, is defined as the value that maximizes $f(\mathbf{P}', \chi_0|P)$.

To find $\hat{p}_{\chi_0}$, we start by writing the joint p.d.f. for the observed values (Q, U) with asymmetric uncertainties

$$f(P|Q', U') = \frac{1}{2\pi \sigma_Q \sigma_U \sqrt{1 - \rho^2}} e^{\left(-\frac{1}{2(1-\rho^2)} \left[\frac{(Q-Q')^2}{\sigma_Q^2} + \frac{(U-U')^2}{\sigma_U^2} - \frac{2\rho(Q-Q')(U-U')}{\sigma_Q \sigma_U}\right]\right)}, \quad (2.25)$$

where ρ is the correlation coefficient between (Q, U), defined as

$$\rho = \frac{E[(Q - Q')(U - U')]}{\sigma_Q \sigma_U} = \frac{\sigma_{QU}}{\sigma_Q \sigma_U}. \quad (2.26)$$

Using that $Q = P \cos 2\chi$ and $U = P \sin 2\chi$ in Eq. 2.25,

$$f(P|Q', U') = \frac{1}{2\pi \sigma_Q \sigma_U \sqrt{1 - \rho^2}}$$
$$e^{\left(-\frac{1}{2(1-\rho^2)} \left[\frac{(P\cos 2\chi - Q')^2}{\sigma_Q^2} + \frac{(P\sin 2\chi - U')^2}{\sigma_U^2} - \frac{2\rho(P\cos 2\chi - Q')(P\sin 2\chi - U')}{\sigma_Q \sigma_U}\right]\right)}. \quad (2.27)$$

$\hat{p}_{\chi_0}$ is defined by the condition,

$$\left.\frac{\partial f(P|Q', U')}{\partial P}\right|_{P=\hat{p}_{\chi_0}} = 0. \quad (2.28)$$

This finally leads to the expression for the de-biased polarisation amplitude,

$$\boxed{\hat{p}_{\chi_0} = \frac{\sigma_U^2 Q' \cos 2\chi - \sigma_{QU}(Q' \sin 2\chi + U' \cos 2\chi) + \sigma_Q^2 U' \sin 2\chi}{\sigma_U^2 \cos^2 2\chi - 2\sigma_{QU} \sin 2\chi \cos 2\chi + \sigma_Q^2 \sin^2 2\chi}.} \quad (2.29)$$

We note that if the "observed" position angle $\chi' = 0.5 \arctan(U'/Q')$ differs by more than 45° from $\chi, \hat{p}_{\chi_0}$ will have a negative value. We also note that this estimator is not valid when the observed polarisation angle is used as a surrogate for χ_0. In that case, $\hat{p}_{\chi_0} = p'$ and there is no correction whatsoever. This means that we cannot use this method, for instance, to correct WMAP K-band if we are using the angle information from K-band data as well. However, it can be used to correct Ka, Q, V and W bands using the additional polarisation angle information from K band.

We would also like to have confidence intervals for this estimator. For this, we use the following definitions

$$Q = P \cos 2\chi, \qquad U = P \sin 2\chi$$
$$Q' = \hat{p}_{\chi_0} \cos 2\chi_0, \qquad U' = \hat{p}_{\chi_0} \sin 2\chi_0, \quad (2.30)$$

and with the assumption that we know the *true* value for $\chi = \chi_0$, we obtain

$$Q - Q' = \cos 2\chi (P - \hat{p}_{\chi_0})$$
$$U - U' = \cos 2\chi (P - \hat{p}_{\chi_0}). \tag{2.31}$$

Then, Eq. 2.25 can be written in the form

$$f(P|\bar{p}) \propto e^{\left(-\frac{\sigma_U^2 \cos^2 2\chi (P - \hat{p}_{\chi_0})^2 + \sigma_Q^2 \sin^2 2\chi (P - \hat{p}_{\chi_0})^2 - 2\rho\sigma_Q\sigma_U \cos 2\chi \sin 2\chi (P - \hat{p}_{\chi_0})^2}{2(1-\rho^2)\sigma_Q^2\sigma_U^2}\right)}$$

$$f(P|\bar{p}) \propto e^{-\frac{(P - \hat{p}_{\chi_0})^2}{2\sigma_{\hat{p}}^2}}, \tag{2.32}$$

with

$$\boxed{\sigma_{\hat{p}_{\chi_0}}^2 = \frac{\sigma_Q^2\sigma_U^2 - \sigma_{QU}^2}{\sigma_U^2 \cos^2 2\chi - 2\sigma_{QU} \sin 2\chi \cos 2\chi + \sigma_Q^2 \sin^2 2\chi}.} \tag{2.33}$$

This is the variance of the $\hat{p}_{\chi_0}$ estimator.

2.5.4 Tests of the Estimators

We compared the effectiveness of the de-biasing methods using Monte Carlo simulations for a range of SNR. First, we study the residual bias in a single pixel for a range of SNR in (Q, U). Then we tested the methods in simulations using the *WMAP* covariance matrices.

General case

Using Monte Carlo simulations, we measure the residual bias in a single pixel for three polarisation amplitude estimators:

- $\hat{p} = P' = \sqrt{q'^2 + u'^2}$; i.e. the naive estimator with no correction for the bias.
- $\hat{p}_{wk} = \sqrt{P'^2 - (\sigma_Q^2 \sin^2 2\chi' + \sigma_U^2 \cos^2 2\chi' - 2\sigma_{QU} \cos 2\chi' \sin 2\chi')}$; the generalised Wardle and Kronberg estimator for asymmetric uncertainties from Eq. 2.22.
- The known angle estimator $\hat{p}_{\chi_0}$, presented in Eq. 2.29.

We first create a grid of 100×100 different values of Q/σ_Q and U/σ_U, in the range $0 \leq Q/\sigma_Q < 10$, $0 \leq U/\sigma_U < 10$, using a uniform spacing. The value of σ_Q is fixed to $\sigma_Q = 1$ while σ_U is scaled using a fixed value for the ratio between the (Q, U) uncertainties (if there are no correlations, $\sigma_Q = \sigma_U = \sigma$). Then, 10^5 Gaussian noise realisations are added to the each point in the grid. The noise realisations have a standard deviation of $\sigma = \sigma_U = 1$. By this way, we have $100 \times 100 \times 10^5$ "observed" values for (Q', U') values.

We then calculated the "observed" polarisation amplitude, $P' = \sqrt{Q'^2 + U'^2}$ from the noisy simulations. We applied the estimators to each simulation and we measured the fractional mean bias for each pixel in the grid, $(\langle \hat{p} \rangle - P_0)/P_0$. The first column of Fig. 2.13 shows the fractional bias of the naive estimator P' using four different values for the error ellipse eccentricity e. In all the simulations, $\sigma_{QU} = 0$. This does not affect the generality of the results because the covariance σ_{QU} can always be set to zero by a rotation of the (Q, U) axes. The second column in Fig. 2.13 shows contours of the fractional mean bias after using the generalised Wardle and Kronberg estimator. The biased regions in the SNR plane reduces considerably in comparison with the first column that has no correction.

The third column of Fig. 2.13 shows the residual bias using the $\hat{p}_{\chi_0}$ estimator. This estimator requires an independent value for the polarisation angle along with the observed values (Q', U') and their standard deviation. We generated additional 10^5 Gaussian realisations for the polarisation angle, centred at the true value, χ_0, for each SNR value. The standard deviation in the angle distribution is fixed to $1°9$,[4] which means that we are in an ideal case where we know with high accuracy the true polarisation angle. The residual bias of this estimator is very small, less than 5% over most of the parameter space as can be seen in the four right-hand panels of Fig. 2.13.

In the previous ideal case, when the real polarisation angle is known accurately, the known-angle estimator performs much better than the generalised Wardle and Kronberg. This situation with very small uncertainty in the polarisation angle is not very common, so we also quantified the bias using simulations with a larger uncertainty in the polarisation angle. Figure 2.14 shows contours for the mean bias averaged over 10^5 Gaussian realisations, centred at the true polarisation angle χ_0 for three values of the uncertainty in the polarisation angle $\sigma_\chi = 2°8$, $5°7$ and $9°5$ (these values corresponds to a SNR in polarisation of 10, 5 and 3 respectively for the case where $\sigma_Q = \sigma_U$).

Figure 2.14 shows that the known angle estimator over corrects the polarisation amplitude by a small amount (less than 5%) if the uncertainty in the polarisation angle is lower than $\sim 6°$ (central column). If the uncertainty in the polarisation angle is larger, the negative bias can be as high as 10%. We note that the last row of the figure shows that this over correction can be as bad as -30%. However, the large eccentricity in this case ($e = 0.9$) does not occur at least in *WMAP* data (see Fig. 2.12).

We can see that the known angle estimator works particularly well in the case when the polarisation angle is well known. An uncertainty in the input polarisation angle up to $\sim 6°$ produces, in the worst scenario, a negative bias of 5%. The exact values will depend on the polarisation angle of each particular pixel. In the next Section we test the debiasing methods using the *WMAP* noise properties.

WMAP **case**

In *WMAP* data, the five frequency maps present different SNR. This is due to a combination between the nature of the observed emission (synchrotron has

---

[4]$1°9$ corresponds to a SNR of 15 in the polarisation amplitude in the case where the uncertainties are symmetric (Naghizadeh-Khouei and Clarke 1993).

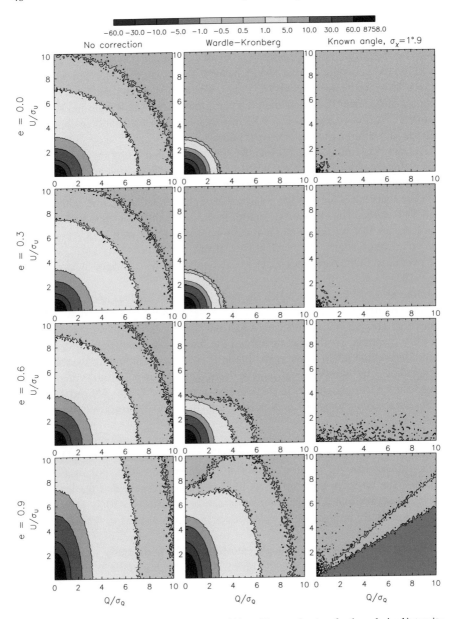

Fig. 2.13 Contours of the fractional mean residual bias of three estimators for the polarised intensity as a function of (Q, U). In each plot, $\sigma_Q = 1$ and $\sigma_{QU} = 0$ (note that there is no loss of generality as the covariance can always be eliminated by a rotation of the $Q - U$ axes). The colour scale represents the percentage bias of the estimated polarisation value for each pixel

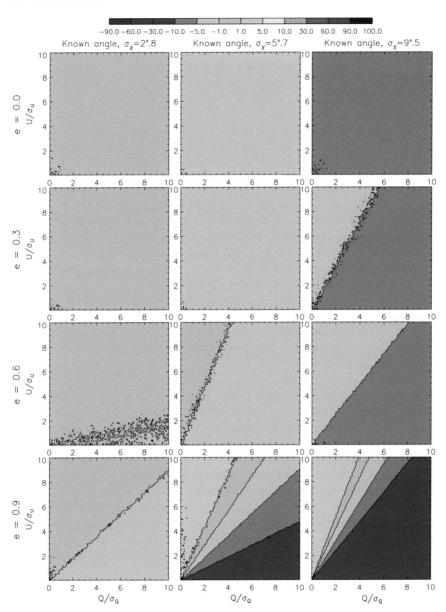

Fig. 2.14 Contours of the mean residual bias of the known angle estimator for three values of the uncertainty in the fiducial polarisation angle (*columns*) and four values for the eccentricity of the error ellipse (*rows*). The colour scale represents the percentage bias of the estimated polarisation value for each pixel

Fig. 2.15 Histograms of the
SNR of the polarisation
probability distribution of
WMAP data. The histograms
are made using the full sky
maps at the original angular
resolution of each map, with
$N_{side} = 512$

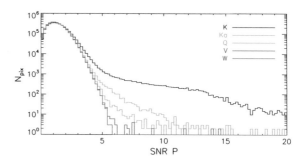

a negative spectral index) and the variation in sensitivity along the bands (see
Table 2.1). *WMAP* K-band has a higher SNR in polarisation, where large areas of
diffuse emission present SNR > 3. Figure 2.15 show histograms of the SNR for
the Stokes parameters (Q, U) of the five frequency bands. The polarisation SNR in
K-band is larger than the SNR in the other bands for almost the entire sky. Also, the
SNR in the Ka-, Q-, V- and W-bands is rarely larger than 3.

We used the *Planck* Sky Model simulation of the polarised sky at K-, Ka- and
Q-bands at an angular resolution of $1°$, from which we can obtain maps for the
unbiased polarisation amplitude, P_0. We added random noise (generated using the
WMAP covariance matrices, in a similar way as described in Sect. 2.3.1) to the simulated Stokes Q and U maps. With these noisy maps, we produce the "observed"
polarisation amplitude, P, maps.

We corrected for the bias in these P maps using both the Wardle and Kronberg's
method and the $\hat{p}_\chi$ estimator described earlier. The angle information required by the
$\hat{p}_\chi$ estimator is measured from the K-band map. This implies that we cannot use the
$\hat{p}_\chi$ estimator to correct the K-band map (it is exactly the same as doing no correction
whatsoever and $\hat{p}_\chi = P$ as mentioned in Sect. 2.5.3). We then compared these de-
biased maps with the true polarisation amplitude map from the PSM simulations. In
Table 2.2 we list the bias value, averaged over the entire sky for the three bands that
we studied.

The absolute mean bias, $\langle \Delta p_0 \rangle = \langle P - P_0 \rangle$ (no correction applied) is larger as
the frequency increases towards Q-band due to the decrease in SNR. For Ka- and
Q-bands, the p_χ estimator performs much better than p_{WK}. We can see this better

Table 2.2 Full-sky averaged values for the polarisation bias, for three estimators

Estimator	K-band μK	Ka-band μK	Q-band μK
$\langle \Delta p_0 \rangle$	0.9 ± 0.1	2.5 ± 0.1	3.00 ± 0.1
$\langle \Delta p_{WK} \rangle$	0.4 ± 0.1	2.0 ± 0.1	2.7 ± 0.1
$\langle \Delta p_\chi \rangle$	0.9 ± 0.1	-0.2 ± 0.1	-0.1 ± 0.1

The first row list the "absolute bias", i.e. the difference between the noisy polarisation amplitude
and the true value for P. The second row lists the residual bias, after using the Wardle and Kroberg's
method. The third row list the averaged bias values for the $\hat{p}_{\chi_0}$ estimator. Note that in K-band,
the $\hat{p}_{\chi_0}$ estimator is not correcting for the bias and that is why the estimated value is equal to the
uncorrected value in the first row

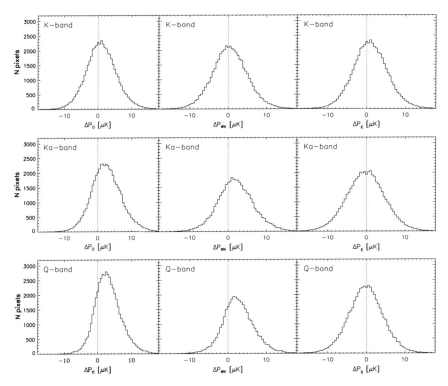

Fig. 2.16 Histograms showing the bias of the three estimators studied. A simulation of the polarised sky at K-, Ka- and Q-band using the Planck sky model is used. The noise present in *WMAP* data is added to each one of the simulated Q, U maps. The histograms show the distribution of the difference between the three estimators and the true polarisation amplitude P'. The column on the *left* shows the histograms of the absolute bias, i.e. the $\Delta P_0 = P_0 - P'$ (no bias correction). The central column shows ΔP using the Wardle and Kronberg estimator, $\Delta P_{WK} = p_{WK} - P'$. For the noisier bands (*Ka* and *Q*), the histograms are not centred at zero, implying that there is additional residual bias on the corrected map. On the *right* the histograms produced using the $\hat{p}_\chi$ estimator, $\Delta P_{WK} = p_\chi - P'$. Here the bias correction works much better and the distributions for the three bands are centred at zero. Table 2.2 lists the central values and uncertainties for all the histograms

by looking at the histograms in Fig. 2.16. In Ka- and Q-band, the bias is clear, the histograms of the uncorrected polarisation maps are not centred at zero (the three plots on the *left*). The Wardle and Kronberg estimator, p_{WK} in the central column, is less biased but in Ka and Q bands there is still a clear shift from the zero value. The p_χ estimator on the right column corrects for the bias much better at the three frequencies, as the three distributions are centred at zero.

In order to see where the residual bias is more important, we show maps of the fractional bias after the correction using the different estimators. Figure 2.17 shows the fractional bias at K-band for the naive estimator $P = \sqrt{Q^2 + U^2}$ and the Wardle and Kronberg estimator $\hat{p}_{WK}$ defined in Eq. 2.22 (the map with the residual bias from the $\hat{p}_{\chi_0}$ estimator is not shown as in this case is equal to the uncorrected estimator).

Fractional bias at K−band using no correction

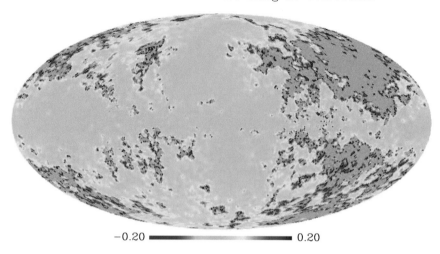

−0.20 ■■■■■■■■■■■■■■ 0.20

Fractional bias at K−band using the WK estimator

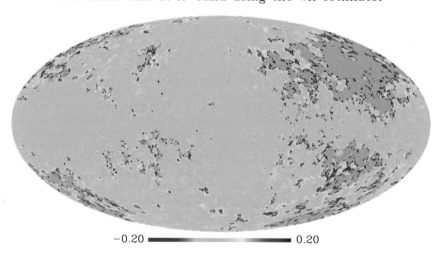

−0.20 ■■■■■■■■■■■■■■ 0.20

Fig. 2.17 Maps showing the fractional bias at K-band. On *top* is the polarisation bias for the naive estimator $P = \sqrt{Q^2 + U^2}$. At the *bottom* is the residual bias that remains after correcting with the $\hat{p}_{WK}$ estimator. All the pixels with a value larger than 20 % (0.2 in the map units) are shown in *grey*. They corresponds to the regions with lower SNR

In the Figure, the pixels where the residual bias is larger than 20 % are shown in grey. The $\hat{p}_{WK}$ estimator leaves a small residual bias over most of the sky (green areas in the Figure).

Figures 2.18 and 2.19 show similar maps for Ka and Q-bands. Here, the residual fractional bias map from the $\hat{p}_{\chi_0}$ estimator have been included. The pixels with an absolute value of the residual bias larger than 20 % have also been masked as grey.

Fractional bias at Ka−band using no correction

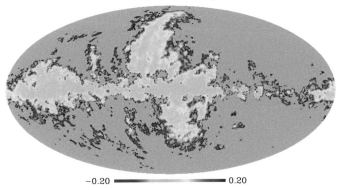

−0.20 ▬▬▬▬▬▬▬ 0.20

Fractional bias at Ka−band using the WK estimator

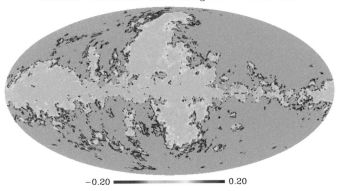

−0.20 ▬▬▬▬▬▬▬ 0.20

Fractional bias at Ka−band using the known−angle estimator

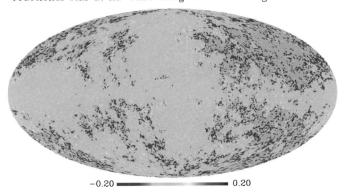

−0.20 ▬▬▬▬▬▬▬ 0.20

Fig. 2.18 Maps showing the fractional bias at Ka-band. On *top* is the polarisation bias for the naive estimator $P = \sqrt{Q^2 + U^2}$. At the *centre* is the residual bias that remains after correcting with the $\hat{p}_{WK}$ estimator. At the *bottom* is the residual bias that remains after correcting with the $\hat{p}_\chi$ estimator. All the pixels with an absolute value larger than 20 % (0.2 in the map units) are shown in *grey*

Fractional bias at Q–band using no correction

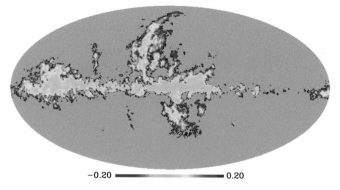

$$-0.20 \quad\rule{2cm}{2pt}\quad 0.20$$

Fractional bias at Q–band using the WK estimator

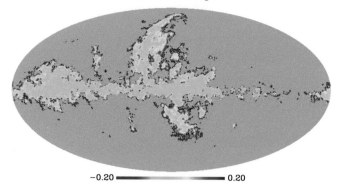

$$-0.20 \quad\rule{2cm}{2pt}\quad 0.20$$

Fractional bias at Q–band using the known–angle estimator

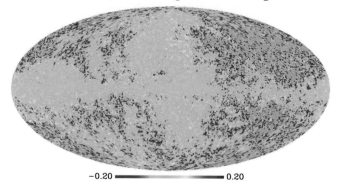

$$-0.20 \quad\rule{2cm}{2pt}\quad 0.20$$

Fig. 2.19 Same as Fig. 2.18 but for *WMAP* Q-band

Table 2.3 Percentage of the area of the full-sky map that have an absolute value of the residual fractional bias smaller than 0.2

Estimator	K-band (%)	Ka-band (%)	Q-band (%)
p_0	84.9	35.0	18.4
p_{WK}	94.7	68.0	42.9
p_χ	–	84.8	84.3

These areas corresponds to the coloured pixels in Figs. 2.17, 2.18 and 2.19

By looking at the unmasked (not grey) areas of Figs. 2.18 and 2.19, we can see that the $\hat{p}_\chi$ estimator performs better than the $\hat{p}_{WK}$ one, as there are much more pixels within the $[-0.2, 0.2]$ range in fractional bias. In Table 2.3 we lists the percentage of the area of the sky with a residual fractional bias smaller than ± 0.2.

2.5.5 Bias-Corrected Polarisation Maps

We prepared bias-corrected polarisation amplitude maps of *WMAP* K, Ka and Q-bands maps, smoothed to a common resolution of 3°. For K-band, we used the modified Wardle and Kronberg formula shown in Eq. 2.22 (Fig. 2.20).

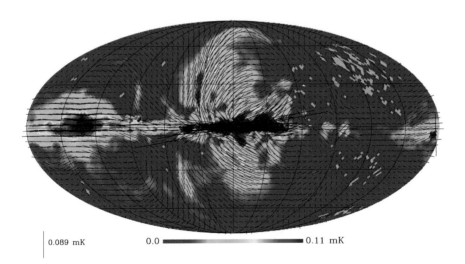

0.089 mK   0.0 ▬▬▬▬▬▬▬▬▬▬ 0.11 mK

Fig. 2.20 Bias-corrected polarisation amplitude maps of *WMAP* K, Ka and Q-bands. The masked area (*grey pixels*) corresponds to the regions where the uncertainty in the polarisation amplitude is larger than the polarisation amplitude so the estimator value is imaginary (see Eq. 2.22). The maps have an angular resolution 3° and the units are thermodynamic Kelvin. The length of the vectors are proportional to the polarisation amplitude (see scale) and their direction is parallel to the magnetic field

Given the nature of the de-biasing method we use for Ka and Q bands—the knowledge of the true polarisation angle—we have masked out the pixels from the K-band map where the uncertainties in (Q, U) would convert to an uncertainty in the polarisation angle, χ, larger than $6°$. This value was chosen to be sure that the residual bias will be smaller than 5 %, as shown with the simulations in the previous Section.

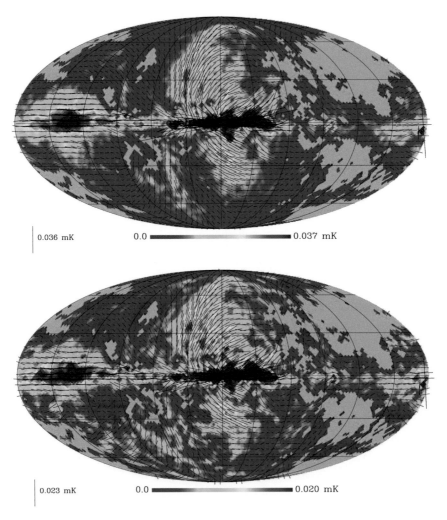

Fig. 2.21 Bias-corrected polarisation amplitude maps of *WMAP* K, Ka and Q-bands. The masked area (grey pixels) corresponds to the region where the uncertainty in the polarisation angle at K-band is larger than $9°$. The maps have an angular resolution $3°$ and the units are thermodynamic Kelvin. The length of the vectors are proportional to the polarisation amplitude (see scale) and their direction is parallel to the magnetic field

This corresponds to a masked area of 17.7 %. In Fig. 2.21 we show the maps, with the masked areas shown as grey. Also plotted on each map are the polarisation vectors, aligned parallel to the magnetic field direction (we have rotated the vectors by 90°).

2.6   Polarisation Upper Limits from AME Regions

Here we summarize the work presented in Dickinson et al. (2011). We used a different de-biasing method to derive upper limits for the fractional polarisation of AME in two molecular clouds. The method used here is a generalisation of the one developed by Vaillancourt (2006) which using Bayesian analysis to estimate the true polarisation amplitude. We modified the method to allow for assymetric uncertainties.

The ρ Ophiuchi and Perseus molecular clouds are two of the most conspicuous AME regions in the Galaxy. Planck Collaboration et al. (2011) describes an analysis in total intensity and showed that these regions are dominated by AME in the 20–60 GHz frequency range. The temperature spectra observed by *Planck* can be fitted using spinning dust models. Here, we used *WMAP* seven-year data to constrain the AME polarisation fraction in these clouds.

We smoothed the maps to a common resolution of $1°$ using the procedure described in Sect. 2.3. Aperture photometry was used to extract flux densities from the *WMAP* maps. We chose a circular aperture of $2°$ to calculate the fluxes. A background level was subtracted by estimating the median value inside an annulus with inner radius equal to $80'$ and an outer radius of $120'$. Figures 2.22 and 2.23 show the I, Q, U, P and polarisation angle maps at the *WMAP* bands for the Perseus and ρ Ophiuchi regions respectively. The integration regions used for the photometry are also shown in the Figures.

The ρ Ophiuchi and Perseus regions are bright sources in intensity maps. Because of this, it is important to test for any potential leakage from Stokes I to Q, U. The *WMAP* team corrects the maps for this leakage. A map of spurious signal, the S map, is calculated using the fact that the intensity leakage does not depend on the orientation of the spacecraft, it only depends on the intensity and spectral shape of the intensity source (Barnes et al. 2003). We measured the polarised signal of the Orion Nebula (M42), a bright un-polarised source to check the leakage level. It has a value $\lesssim 0.1$ % at K-band, and it is even smaller at the higher frequencies. We therefore do not include an additional contribution to the uncertainty in polarisation due to leakage.

The noise in the smoothed Q, U maps was derived using Monte Carlo simulations, which include the correlated terms between Q and U. Then, the polarisation flux densities were corrected for the polarisation bias. We list in Table 2.4 the flux densities for the ρ Ophiuchi and Perseus clouds in all the *WMAP* bands for Stokes I, Q, U and polarisation amplitude. The de-biased polarisation amplitude is very low in both clouds (less than 3σ at all frequencies). We can only place upper limits for the polarisation fraction giving the noise of the data. The limits for the fractional

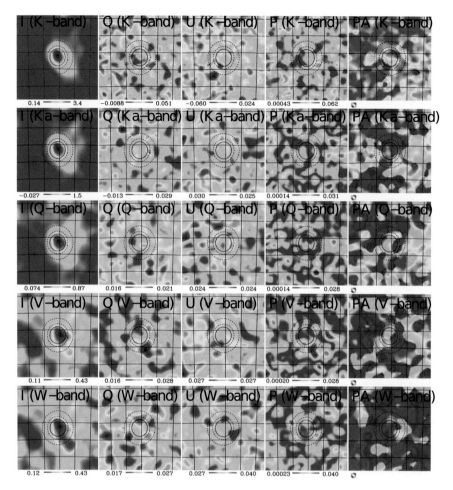

Fig. 2.22 *WMAP* 7-year maps of the Perseus region at the five *WMAP* bands. Each map covers $10° \times 10°$ centred at $(l, b) = (353°.05, +16°.90)$ and is smoothed to $1°$ resolution. From *left* to *right* are Stokes I (total intensity), Q, U, polarized intensity (P) and polarization angle (PA). Units are thermodynamic (CMB) mK. The graticule has a spacing of $2°$. The primary extraction aperture is shown as a *solid line* and the background annulus as a *dashed line*

polarisation of AME in ρ Ophiuchi are 1.7, 1.6 and 2.6 % while in the Perseus clouds are 1.4, 1.9 and 4.7 % at K, Ka and Q bands respectively.

These limits rule out some emission mechanism for the AME. Magnetic dipole radiation can exhibit high ($\sim > 10$ %) polarization at 20–40 GHz when there is a single magnetic domain (Draine and Lazarian 1999). The polarisation fraction limits we found are not compatible with this model.

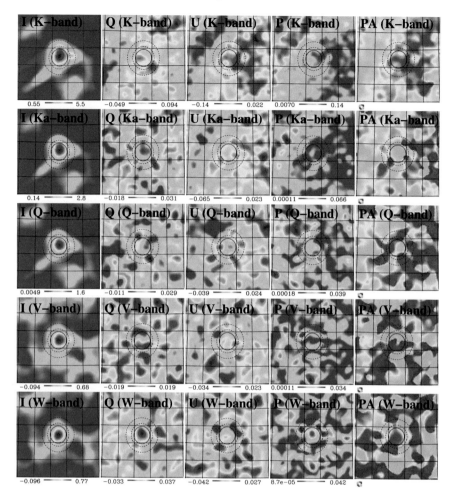

Fig. 2.23 *WMAP* 7-year maps of the ρ Ophiuchi region at the five *WMAP* bands. Each map covers $10° \times 10°$ centred at $(l, b) = (353°.05, +16°.90)$ and is smoothed to $1°$ resolution. From *left* to *right* are Stokes I (total intensity), Q, U, polarized intensity (P) and polarization angle (PA). Units are thermodynamic (CMB) mK. The graticule has a spacing of $2°$. The primary extraction aperture is shown as a *solid line* and the background annulus as a *dashed line*

If these limits of polarisation fraction are typical for AME at high Galactic latitudes, the only important polarisation emission at frequencies under 100 GHz would be synchrotron, therefore simplifying the CMB polarised foreground characterisation.

Table 2.4 Aperture photometry of the *WMAP* 7-year maps in the ρ Ophiuchi (*top*) and Perseus (*bottom*) molecular cloud regions, using a 2° diameter aperture

Frequency (GHz)	I (Jy)	I_{sd} (Jy)	Q (Jy)	U (Jy)	P (Jy)	P_0 (Jy)	Π (%)	Π_{sd} (%)
22.7 (K-band)	26.3 ± 5.5	24.8 ± 6.6	0.091 ± 0.096	0.23 ± 0.14	0.25 ± 0.13	0.21 ± 0.11(<0.43)	<1.6	<1.7
33.0 (Ka-band)	30.7 ± 5.3	27.2 ± 6.3	−0.27 ± 0.12	0.02 ± 0.15	0.27 ± 0.12	0.24 ± 0.12(<0.44)	<1.4	<1.6
40.7 (Q-band)	27.7 ± 4.6	21.9 ± 5.8	−0.07 ± 0.18	−0.23 ± 0.24	0.24 ± 0.23	0.00 ± 0.30(<0.57)	<2.1	<2.6
60.6 (V-band)	26.3 ± 4.5	9.8 ± 6.5	0.35 ± 0.41	0.41 ± 0.40	0.54 ± 0.41	0.00 ± 0.61(<1.1)	<4.2	<11
93.4 (W-band)	63.6 ± 8.9	6 ± 15	2.8 ± 1.0	0.0 ± 1.7	2.8 ± 1.0	2.6 ± 1.0	4.1 ± 1.8	...
22.7 (K-band)	21.0 ± 3.1	16.7 ± 3.5	−0.118 ± 0.071	0.068 ± 0.069	0.136 ± 0.070	0.11 ± 0.06(< 0.24)	<1.1	<1.4
33.0 (Ka-band)	20.4 ± 3.0	15.7 ± 3.3	0.02 ± 0.13	−0.06 ± 0.15	0.07 ± 0.15	0.0 ± 0.15(<0.30)	<1.5	<1.9
40.7 (Q-band)	16.9 ± 2.9	11.6 ± 3.3	0.07 ± 0.19	−0.25 ± 0.21	0.26 ± 0.21	0.0 ± 0.30(<0.54)	<3.2	<4.7
60.6 (V-band)	14.9 ± 4.0	5.4 ± 4.4	0.098 ± 0.077	−0.64 ± 0.59	0.65 ± 0.39	0.4 ± 0.27(<1.2)	<8.1	<22
93.4 (W-band)	32.4 ± 9.8	3 ± 10	−0.5 ± 1.3	−0.3 ± 2.3	0.6 ± 1.4	0.0 ± 1.5(< 2.9)	<9.0	...

The columns contain flux densities for intensity I, the spinning dust contribution to the intensity (I_{sd}), Q, U, the observed polarized intensity (P), the noise-bias corrected polarized intensity (P_0), the total fraction (Π), and the polarization fraction for spinning dust (Π_{sd}). Upper limits are at the 95 % c.l

References

Barnes, C., et al. (2003). First-Year Wilkinson Microwave Anisotropy Probe (WMAP) observations: Galactic signal contamination from sidelobe pickup. *ApJS, 148*, 51–62.

Bennett, C. L., et al. (2013). Nine-year Wilkinson Microwave Anisotropy Probe (WMAP) observations: Final maps and results. *ApJS, 208*(20), 20.

Delabrouille, J., et al. (2013). The pre-launch Planck SkyModel: a model of sky emission at submillimetre to centimetre wavelengths. *A & A, 553*(A96), A96.

Dickinson, C., Peel, M., & Vidal, M. (2011). New constraints on the polarization of anomalous microwave emission in nearby molecular clouds. *MNRAS, 418*, L35–L39.

Draine, B. T., & Lazarian, A. (1999). Magnetic dipole microwave emission from dust grains. *ApJ, 512*, 740–754.

Fixsen, D. J. (2009). The temperature of the cosmic microwave background. *ApJ, 707*, 916–920.

Górski, K. M., et al. (2005). HEALPix: A framework for high-resolution discretization and fast analysis of data distributed on the sphere. *ApJ, 622*, 759–771.

Haslam, C. G. T., et al. (1982). A 408 MHz all-sky continuum survey. II—The atlas of contour maps. *A & AS, 47* 1.

Hinshaw, G., et al. (2003). First-Year Wilkinson Microwave Anisotropy Probe (WMAP) observations: Data processing methods and systematic error limits. *ApJS, 148*, 63–95.

Jarosik, N., et al. (2003). First-Year Wilkinson Microwave Anisotropy Probe (WMAP) observations: On-orbit radiometer characterization. *ApJS, 148*, 29–37.

Mather, J. C., et al. (1999). Calibrator design for the COBE Far-Infrared Absolute Spectrophotometer (FIRAS). *ApJ, 512*, 511–520.

Miville-Deschênes, M.-A., et al. (2008). Separation of anomalous and synchrotron emissions using WMAP polarization data. *A & A, 490*, 1093–1102.

Naghizadeh-Khouei, J., & Clarke, D. (1993). On the statistical behaviour of the position angle of linear polarization. *A & A, 274*, 968.

Planck Collaboration et al. (2011). Planck early results. XX. New light on anomalous microwave emission from spinning dust grains. *A & A, 536*, A20, A20.

Quinn, J. L. (2012). Bayesian analysis of polarization measurements. *A & A, 538*, A65, A65.

Serkowski, K. (1958). Statistical analysis of the polarization and reddening of the double cluster in perseus. *Acta Astron, 8*, 135.

Simmons, J. F. L., & Stewart, B. G. (1985). Point and interval estimation of the true unbiased degree of linear polarization in the presence of low signal-to-noise ratios. *A & A, 142*, 100–106.

Vaillancourt, J. E. (2006). Placing confidence limits on polarization measurements. *PASP, 118*, 1340–1343.

Vinokur, M. (1965). Optimisation dans la recherche d'une sinusode de période connue en présence de bruit. Application à la radioastronomie. *Annales d'Astrophysique, 28*, 412.

Wardle, J. F. C., & Kronberg, P. P. (1974). The linear polarization of quasi-stellar radio sources at 3.71 and 11.1 centimeters. *ApJ, 194*, 249–255.

Wehus, I. K., Fuskeland, U., & Eriksen, H. K. (2013). The effect of systematics on polarized spectral indices. *ApJ, 763*(138), 138.

Chapter 3
WMAP Polarised Filaments

At frequencies up to a few GHz, the total radio sky brightness is dominated by the diffuse Galactic synchrotron radiation, where cosmic rays (CR) electrons and positrons spiralling around magnetic field lines emit radiation. Synchrotron is the most important emission mechanism at frequencies below ~1 GHz. At higher frequencies, free-free radiation, anomalous microwave emission and the thermal dust radiation dominate the temperature spectrum of the sky (Davies et al. 2006).

The diffuse synchrotron Galactic emission has a smooth component and considerable structure at high Galactic latitude. Some of these structures, the "radio loops" are also visible in microwaves, X-rays and gamma-rays. These loops are some of the largest structures on the sky and have been studied for more than 50 years. Brown et al. (1960) reviewed for the first time possible theories about the origin of the most obvious of them, Loop I, also called the North Polar Spur (NPS). Loop II (Large et al. 1962), Loop III (Quigley and Haslam 1965) and Loop IV (Large et al. 1966) were discovered shortly after. The emission from all these features is non-thermal with spectral indices β around 1 GHz ranging between -2.7 and -2.9 (Berkhuijsen 1973; Borka 2007). These loops are also visible in polarisation, both in starlight and radio (Mathewson and Ford 1970; Page et al. 2007).

Loop I has an HI counterpart, emits soft X-rays (Egger and Aschenbach 1995) and appears to have a limb-brightened perimeter. Heiles (1998) points out that Loops II, III & IV are too diffuse to be limb-brightened structures. There are different interpretations for the origin of these loops:

- Outflow from Galactic Centre (Bland-Hawthorn and Cohen 2003; Sofue 1977)
- Bubbles/shells powered by OB associations (Egger 1995; Wolleben 2007).
- Old and nearby supernova remnants (Berkhuijsen et al. 1971; Spoelstra 1973) and magnetic field loops illuminated by relativistic electrons (Heiles 1998).

In this chapter we will focus the analysis on the filamentary structure observed in polarisation by *WMAP*. We will measure polarisation spectral indices, calculate

© Springer International Publishing Switzerland 2016
M. Vidal Navarro, *Diffuse Radio Foregrounds*, Springer Theses,
DOI 10.1007/978-3-319-26263-5_3

polarisation fraction, study the Faraday rotation, see the connection between the filaments with the local interstellar medium and quantify the contribution of the polarised filamentary emission to the CMB power spectra.

3.1 Polarised Large Scale Features

As previously noted, most of the polarised emission at high latitudes comes from individual filamentary features. Some of these structures are the well-known continuum radio loops. Other filaments only appear in polarisation and we will describe them in this section. Some of these structures are very large in the sky and can be the dominant CMB foreground in polarisation.

3.1.1 Continuum Loops in Polarisation

A number of radio loops have been described in the literature, Loop I or NPS being the best known. These loops are well-fitted by small circles on the sky (see e.g. Berkhuijsen et al. 1971). Most of them have been observed in low frequency ($\nu <$ 1 GHz) continuum radio surveys. Six loops have been described in the literature. Table 3.1 lists some properties of the loops and the references. In Fig. 3.1 we use the 45 MHz map from Guzmán et al. (2011) to show the location of the four best known continuum loops.

We use the *WMAP* K-band polarisation map with the unsharp mask applied (see Sect. 2.4) to identify the circular features. In Fig. 3.2 we show this map, which has been filtered using a Gaussian smoothing kernel of 10° FWHM. A number of filaments are easily recognisable over the sky, most of them lie in the inner Galaxy, with Galactic longitude in the range $-90° < l < 90°$. Here we concentrate on the filaments that have a "circular" arc shape. The continuum Loops I, III and IV are visible in polarisation in this filtered map. The more diffuse Loop II is not obvious on this map. It is interesting that the polarisation map, after filtering, preserves most of the emission (compare the filtered map in Fig. 3.2 with the original K-band polarisation amplitude map shown in Fig. 2.4). In other words, most of the polarised emission originates in these filamentary structures.

We fitted small circle arcs to the coordinates of the pixels that trace the peak of the circular features from Fig. 3.2. Geometrically, a small circle intercepts the sphere on a plane, which is normal to the vector that defines the coordinates of the centre of the circle. This plane can be described in Cartesian coordinates using the vector equation,

$$\mathbf{z} = a + b\,\mathbf{x} + c\,\mathbf{y}, \tag{3.1}$$

where (a, b, c) are constants.

Table 3.1 Circular loops and arcs parameters

Loop name	Continuum			Polarisation			Comments/References
	l	b	r	l	b	r	
I	329.0	17.5	58	332.6	20.7	54.3	NPS, Brown et al. (1960)
II	100.0	−32.5	45.5	–	–	–	Large et al. (1962)
III	124.0	15.5	32.5	118.8	13.2	31.6	Quigley and Haslam (1965)
IV	315.0	48.5	19.8	315.8	48.1	19.3	Large et al. (1966)
V	127.5	18.0	67.2	–	–	–	Milogradov-Turin and Uroišević (1997), same as loop III
VI	120.5	30.8	72.4	–	–	–	Milogradov-Turin and Uroišević (1997), not visible in *WMAP* polarisation
VIIa	–	–	–	345.0	0.0	65	From Wolleben (2007) using the 1.4 GHz map
VIIb*				0.7	−23.3	45.9	Values found in this work for VIIa
VIII	–	–	–	344.0	4.8	18.5	Includes the GC loop from Sofue et al. (1989)
IX*	–	–	–	332.0	16	46.5	Inside the NPS
X*	–	–	–	106	−22	50.0	Below the plane at the "fan" region
XI*	–	–	–	30	35	67.0	Tangential to the plane
XII*	–	–	–	227	38	81.0	Large arc tangential to the plane
XIII	–	–	–	300	0.7	27.6	Close to the LMC

The *top* half lists the previously identified continuum loops. *Bottom* half lists the loops and circular arcs that are visible only in polarisation. The 11 loops and arcs that are visible in polarisation are shown on the *bottom* panel of Fig. 3.2. All the numerical values on the table are in degrees
*These arcs are clearly visible only in polarisation

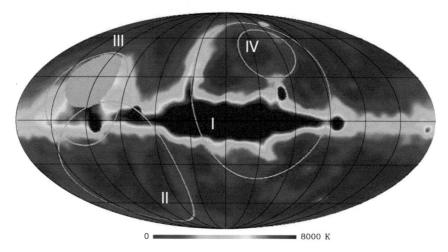

Fig. 3.1 45 MHz map of Guzmán et al. (2011) showing the four best known continuum loops

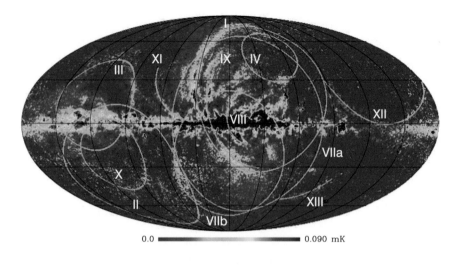

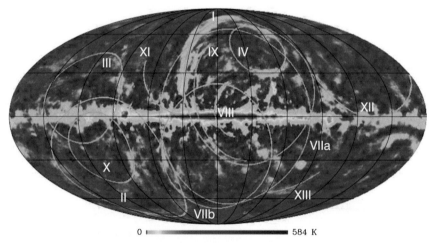

Fig. 3.2 *Top* Un-sharp mask version of the *WMAP K* band polarisation amplitude map. The angular resolution of the unfiltered maps is 1° and the filter beam has a size of 10° FWHM. The location and sizes of the 13 loops and arcs are listed in Table 3.1. We use the polarisation template shown on *top* to identify polarised circular features that are shown in the figure. *Bottom* Un-sharp mask version of the Haslam et al. (1982) map

In order to find the centre of the arc, we first converted the Galactic coordinates (l, b) of each point that belongs to the polarisation loop to Cartesian coordinates $(\mathbf{x}, \mathbf{y}, \mathbf{z})$. We then used a least-squares fit to find the plane represented by Eq. 3.1 that describes the data better. We finally calculated the radius by taking the median of the angular distance from the centre to the points that belong to the loop.

Table 3.1 lists the parameters found for the three previously known continuum loops that are visible in the polarisation data, as well as the parameters published on the literature from continuum fits to the data. We did not attempt to fit for Loop II (Cetus Arc) due to the very low emission observed in the *WMAP* polarisation maps. There is a good agreement for the radii and centres of the loops between the low frequency continuum data and the *WMAP* polarisation data. Loop III is the one that shows the larger discrepancy where the polarisation peaks around 5° away from the continuum. This shift between the total intensity and the polarisation morphology was noted earlier by Spoelstra (1972).

There are a number of other features visible in K-band polarisation that can be fitted by small circle arcs. We have identified five new arcs that are visible only in these polarisation data (marked with an asterisk in Table 3.1).

Wolleben (2007) identifies a "New Loop" (loop VIIa in our nomenclature) using the 1.4 GHz polarisation map from the DRAO survey (Wolleben et al. 2006). The 1.4 GHz maps present a depolarisation band at Galactic latitudes $|b| < 30°$ due to Faraday depolarisation (see Sect. 3.4). Because of this, they only use data at $b < -35$ to define the location and size of this new loop. Here we include data that is closer to the Galactic plane that belongs to the arc in order to define its geometry. The filament is brighter closer to the Galactic plane, so we think that our definition for the geometry of this new loop is more appropriate (in Fig. 3.2, compare the loops VIIa and VIIb).

3.2 Polarisation Angle Along Filaments

The direction of the ordered component of the magnetic field can give us information about the origin of the observed filaments. The magnetic fields of old SN (e.g. the DA 530 supernova, Landecker et al. 1999) are tangential to the surface of the shell. This is traditionally explained by the compression of the field in radiative shocks with large radius (see e.g. the review by Reynoso et al. 2012). On the other hand, younger SN can show radial magnetic fields close to the border of the shell, as in Tycho's SN (Reynoso et al. 1997). The origin of the radial fields is not entirely clear and MHD simulations are used to understand this effect. For instance, Jun and Jones (1999) showed that the turbulent interaction of the expanding shock with the ISM can amplify the radial component of the magnetic field.

Looking at the de-biased *WMAP* K-band polarisation map from Fig. 2.21, we see that most of the emission at high latitudes comes from individual filamentary structures. We also can see that the magnetic field direction vectors are roughly parallel to the direction of the filaments. We will quantify this observation by comparing the polarisation angle χ with the direction of the filament.

In the *top* panel of Fig. 3.2, there are some features that are not well fitted by a circular arc. We include some of these non-circular features in this analysis, and also ignore some of the loops and arcs that are listed in Table 3.1 due to their low brightness. The filaments that we use here are marked with blue crosses in the top panel of Fig. 3.3.

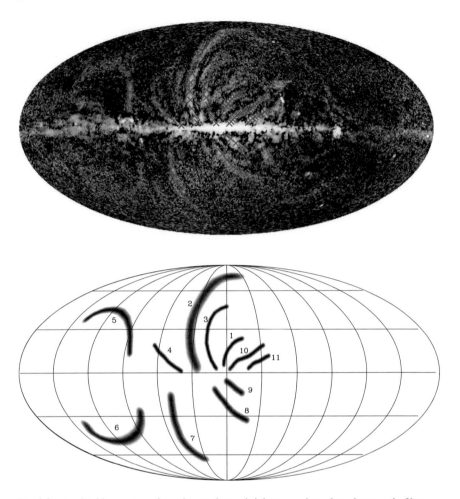

Fig. 3.3 *Top* the *blue crosses* show the maximum brightness points along large-scale filaments visible on the unsharp-mask version of the *WMAP* K-band polarisation amplitude map. *Bottom* template map showing the filaments visible on the *top* panel

In order to define the direction α of the filaments that we want to compare with the polarisation angle χ, we connected pairwise the points of maximum brightness across each filament shown in Fig. 3.3 using great circle arcs. This created a map with a smooth line that follows the direction of each filament. Then, this maximum brightness template for each filament was convolved with a Gaussian beam of a width similar to the apparent width of the filament, typically $2°$–$4°$. In the bottom panel of Fig. 3.3 we can see the resulting templates for all the filaments that we study here.

The direction at each point of the filament, α, can be calculated using the spatial gradient of the previously described template map for the filaments, T. The gradient of this map points to a direction that is perpendicular to the extension of the filament, therefore we will have,

$$\alpha = \arctan\left(\frac{\partial T/\partial\theta}{\partial T/\partial\phi/\sin\theta}\right), \tag{3.2}$$

where θ and ϕ are the spherical coordinates. The $\sin(\theta)$ factor appears because of the use of spherical coordinates.

As we are interested in the comparison with the polarisation angle χ, we impose $\alpha = \alpha \pm \pi$, as the polarisation angle is invariant with $\pm\pi$. Finally, we subtract $\pi/2$ from alpha to make it parallel to the direction of the filament. Figure 3.4 shows two examples of the maps that define the angle α, for filaments No. 3 and for the NPS.

We measured the difference between the observed polarisation angle χ as defined in Eq. 1.5 and the direction angle α. We did this in adjacent circular apertures along each filament. Figures 3.5, 3.6, 3.7 and 3.8 show on the left a map of each filament with the circular apertures where we calculate the angle difference. A thin grey line indicates the direction of the filament defined from the template map and the gradient method described. The plots on the right show the difference between the observed polarisation angle and the angle of the filament. There are two error bars, the shorter (red) represents the random fluctuations given by the dispersion of the measured angles in aperture. The longer (blue) bars include an additional 5° systematic uncertainty added to account for small errors in the definition of the spatial direction of each filament.

For the NPS (filament number 2 in our numbering), we also compared the polarisation angle χ with the direction angle defined by the geometry of the loop in total intensity, i.e. the circular fit that traditionally is used to describe the NPS, listed in Table 3.1. The comparison can be seen in the lower panel of Fig. 3.5. The green line in that figure represents the angle that the circular loop defines. The difference between the circular loop and our definition for the direction of the NPS using the gradient technique is small compared with the uncertainties, so there is consistency between the two definitions for the direction of the filament.

In Filament No. 1, the difference in angle is small, averaging less that 5°. The magnetic field vectors are very close to parallel with the direction of the filament. In

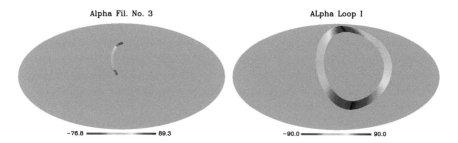

Fig. 3.4 Two examples of maps of the direction angle, α of filament 3 (*left*) and the NPS (*right*)

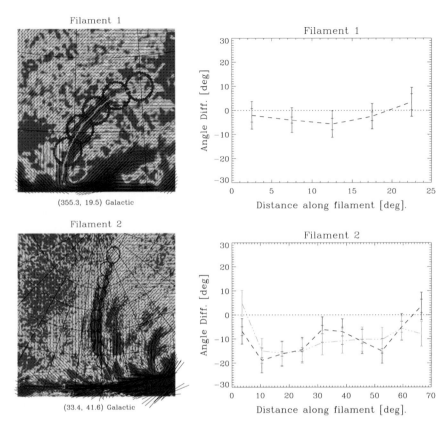

Fig. 3.5 The panels on the *left* show maps of each filament as defined in Fig. 3.3. The adjacent *circles* in each filament show the apertures that we use to measure the difference between the polarisation angle (represented by the small *black* vectors) and the direction of the filament (defined by *thin grey line* running along the filament). The panels on the *right* show the difference between the polarisation angle, χ, and the angle defined by the direction of each filament, α, along the filament. In these plots, the distance along the filament is measured always from the extreme closer to the Galactic plane. The *red error bars* represent the random fluctuation in the polarisation angle inside the aperture. The *blue bar* includes the random fluctuations and a 5° uncertainty, assigned as a conservative systematic error in the definition of the direction of the filament

Filament No. 2, the NPS, there is a negative offset of ∼10° in the angles. It can be seen even by looking at the map on Fig. 3.5. The same negative difference occurs in filament No. 3. Filament No. 4 is also well-aligned, with a difference very close to zero within the errors. Filaments No. 5, 6 and 7 are those with less signal-to-noise ratio, and the polarisation angle on them is not parallel to their extension, there is even an inversion in the sign of the difference of the angles. Filaments No. 8 and 9 also have a systematic difference in angles but in this case, the difference is positive. Emission from Filament No. 10 is well aligned along its extension. Finally, filament No. 11 shows a systematic positive difference of ∼10°.

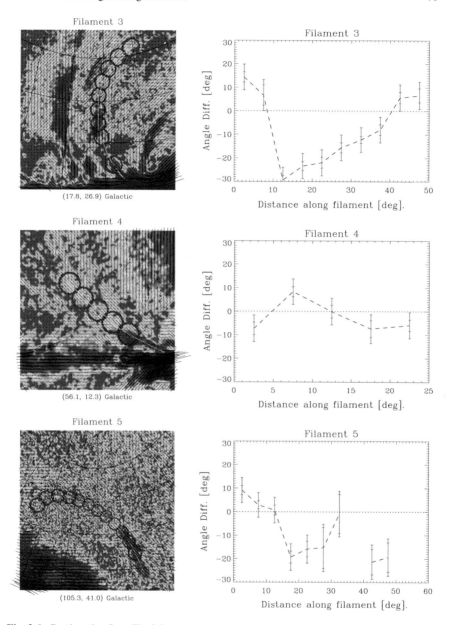

Fig. 3.6 Continuation from Fig. 3.5

It is interesting that in the inner Galaxy, the deviations from parallel in the direction of the angles are different on the north Galactic hemisphere compared with the south (compare filaments Nos. 2 and 3 in the north with filaments Nos. 8 and 9 in the south). The direction of the emission from these filaments has a smaller radius of

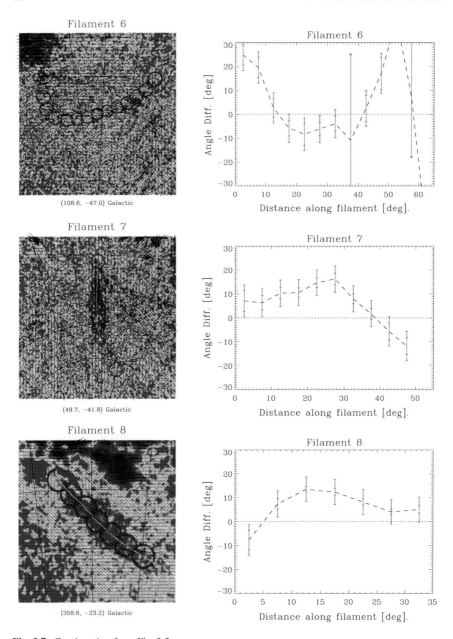

Fig. 3.7 Continuation from Fig. 3.5

curvature than the one defined by the polarisation vectors. Filament No. 1, on the other hand is particularly well aligned with the polarisation vectors. We will discuss these observations later in the context of the origin of the filamentary emission.

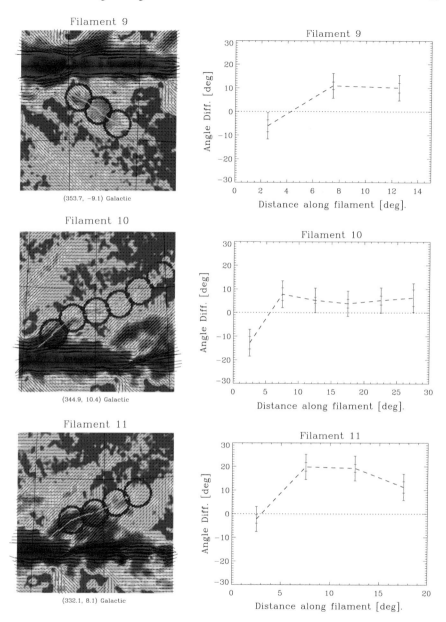

Fig. 3.8 Continuation from Fig. 3.5

We note that in each case when there is a change in the sign of the $\chi - \alpha$, there is an overlapping feature. An example of this can be seen in Filament No. 10 from Fig. 3.8, where the first circular aperture closer to the Galactic plane includes emission both from the filament and from the plane.

3.3 Spectral Indices

Temperature spectral indices are useful to identify the emission mechanism of the measured radiation and by doing so, obtain physical properties of the emitting region. Because of the combination of different emitting sources along an arbitrary line-of-sight, it is difficult to measure the spectral index of an individual physical emission mechanism. Davies et al. (2006) selected small regions of the sky that were expected to be dominated by a single emission mechanism, which allowed them to measure the spectral indices of free-free, synchrotron and thermal dust radiation.

Where polarisation data are available, the synchrotron spectral index can be measured unambiguously without the use of component separation techniques in the frequency range below $\sim$50 GHz, because the polarisation of the other mechanisms is much smaller ($\lesssim$1 %). The measurement of precise synchrotron spectral indices has become even more necessary after the discovery of the "haze" by Finkbeiner (2004), a diffuse emission of unknown origin centred at the Galactic centre. Planck et al. (2013) has confirmed the existence of the haze and they have measured a spectral index of -2.55 ± 0.05, favouring a hard-spectrum synchrotron origin for this diffuse emission. This confirms that significant synchrotron spectral index variations can occur and they should therefore be measured with care. We note that the frequency-cleaned CMB maps that the *WMAP* team provide are produced on the basis of a fixed spectral index for synchrotron emission.

In this section we measure the spectral index of the bias-corrected polarised intensity between *WMAP* K, Ka and Q bands in a number of regions across the sky. We use the T-T plot approach, in which we assume a power-law relationship between the intensities at different frequencies:

$$\frac{T_{\nu_1}}{T_{\nu_2}} = \left(\frac{\nu_1}{\nu_2}\right)^{\beta}. \tag{3.3}$$

This method has the advantage of being independent of any zero level present in the maps. Here, the spectral index β is calculated by fitting a straight line $y = mx + n$ in logarithmic space to the measured brightness temperatures T_{ν_1}, T_{ν_2}:

$$\beta = \frac{\log m}{\log(\nu_1/\nu_2)}, \tag{3.4}$$

where m is the slope of the fit. We took into account the error in both coordinates for the linear fit. Also, an absolute calibration error of 0.2 % (Bennett et al. 2013) has been added in quadrature to the random uncertainty.

3.3.1 T-T Plots of the Filaments

We calculated the spectral indices in the same regions that encompass the filaments shown in Fig. 3.3. Due to the superposition of emitting regions on the Galactic plane, it is difficult to isolate the emission that comes only from a filament, so we have masked all the pixels with $|b| \leq 6°$. Figures 3.9, 3.10 and 3.11 show the T-T plots for the eleven different filaments. In Table 3.2 are listed the spectral indices, the reduced χ^2 of the fit and the χ^2 probability, q, for each T-T plot.

Most of the filaments show a spectral index consistent with -3.0. Filaments No. 4 and No. 9 have a β_{K-Ka} flatter than -3.0 with 1-σ significance, then these spectra get steeper with increasing frequency. The spectral index of the NPS (Filament No. 2) of $\beta_{K-Ka} = -3.11 \pm 0.27$ is consistent with the value measured using total intensity data ($\beta = -3.07^{+0.09}_{-0.13}$ by Davies et al. 2006). Filament No. 1, the Galactic centre spur, has a steep spectral index between K and Ka bands of $\beta_{K-Ka} = 3.23 \pm 0.24$, which becomes flatter with higher frequency, $\beta_{K-Q} = 2.91 \pm 0.24$ and $\beta_{Ka-Q} = 2.02 \pm 0.76$. The same behaviour happens in filaments No. 3 and No. 7 although in this last one, the uncertainties are large so the flattening of the spectra is not significant.

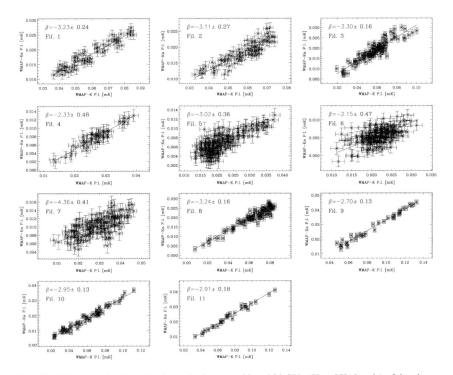

Fig. 3.9 T-T plots of polarisation intensity between 23 and 33 GHz (K and Ka bands) of the eleven filaments defined in Fig. 3.3. The *straight line* shows the best linear fit. The *error bars* show only the statistical fluctuations for each point. The uncertainty in the spectral index β includes the 0.2 % calibration error

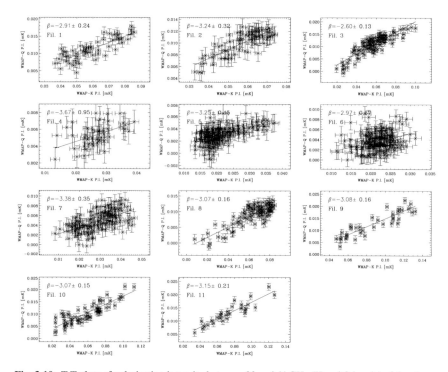

Fig. 3.10   T-T plots of polarisation intensity between 23 and 41 GHz (K and Q bands) of the eleven filaments defined in Fig. 3.3. The *straight line* shows the best linear fit. The *error bars* show only the statistical fluctuations for each point. The uncertainty in the spectral index β includes the 0.2 % calibration error

Some filaments show a large dispersion of the data and two slopes can be seen. This indicates that there is more than one spectral component in the studied region. Examples of this are filaments 2 and 3. We will try to overcome this situation by defining smaller regions on the sky to calculate the spectral indices. This is described in the next section.

3.3.2   T-T Plots on Small Regions

We selected small regions from the filaments and also some areas of interest that do not necessarily belong to a filament, including regions on the Galactic plane. Figure 3.12 shows the regions that were chosen. Most of them lie in the inner Galaxy and three are in the Fan region, around $l = 140°$.

In Figs. 3.13, 3.14 and 3.15 we show the T-T plots for each region. Table 3.3 lists the location and size of each region, the measured spectral indices, the reduced χ^2

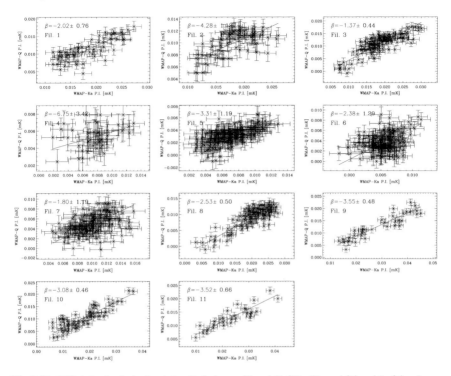

Fig. 3.11 T-T plots of polarisation intensity between 33 and 41 GHz (Ka and Q bands) of the eleven filaments defined in Fig. 3.3. The *straight line* shows the best linear fit. The *error bars* show only the statistical fluctuations for each point. The uncertainty in the spectral index β includes the 0.2 % calibration error

Table 3.2 Spectral indices of the filaments between *WMAP* K, Ka and Q bands

Region	β_{K-Ka}	χ^2_{red}	q	β_{K-Q}	χ^2_{red}	q	β_{Ka-Q}	χ^2_{red}	q
1	-3.33 ± 0.29	0.67	0.96	-2.96 ± 0.27	1.22	0.14	-1.89 ± 0.87	0.87	0.74
2	-3.17 ± 0.27	1.35	0.03	-3.17 ± 0.30	0.84	0.81	-4.10 ± 1.11	1.36	0.03
3	-3.35 ± 0.18	1.35	0.01	-2.71 ± 0.15	1.18	0.11	-1.55 ± 0.52	1.22	0.07
4	-2.50 ± 0.47	0.34	1.00	-3.50 ± 0.78	1.10	0.31	-5.44 ± 2.86	1.23	0.15
5	-3.17 ± 0.31	0.80	0.95	-3.58 ± 0.39	0.63	1.00	-4.12 ± 1.07	0.51	1.00
6	-3.05 ± 0.68	0.97	0.58	-2.81 ± 0.64	1.25	0.03	-0.18 ± 1.39	0.84	0.91
7	-4.22 ± 0.45	1.11	0.20	-3.64 ± 0.47	0.98	0.55	-2.40 ± 1.34	0.88	0.81
8	-3.28 ± 0.20	0.81	0.88	-3.29 ± 0.22	1.27	0.05	-3.04 ± 0.67	0.97	0.56
9	-2.73 ± 0.16	0.68	0.92	-3.13 ± 0.21	1.48	0.04	-3.65 ± 0.61	0.99	0.48
10	-2.95 ± 0.15	0.71	0.95	-3.20 ± 0.18	1.63	0.00	-3.46 ± 0.53	1.36	0.04
11	-2.90 ± 0.19	0.50	0.98	-3.22 ± 0.25	1.14	0.28	-3.75 ± 0.77	1.18	0.24

Also listed are the reduced of the fit and the χ^2 probability or q-value, q, for each T-T plot

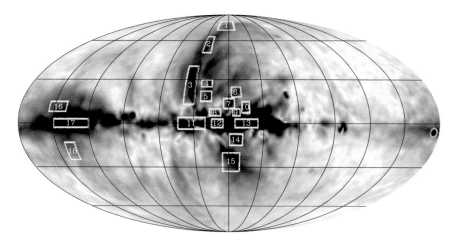

Fig. 3.12 *WMAP* K-band polarisation intensity map showing the 18 smaller regions chosen to measure the spectral indices

value and the q-value. The fits in these smaller regions are in general better than those in Figs. 3.9, 3.10 and 3.11 due to a smaller intrinsic dispersion of the data. The T-T plots between K and Ka bands, where the SNR is larger, allow us to measure β with small uncertainties. This occurs mainly on the Galactic plane, where the signal is stronger. At higher latitudes, the fractional uncertainties on the data are larger so the constraints on β are less tight.

There are a few regions that show a flatter spectral index than the commonly accepted -3.0 value. Regions 1 and 2, at the top of the NPS have $\beta_{K_{K}a} > -3.0$ at a 2-σ significance level. Also, in the Fan region, regions No. 16, No. 17 and No. 18 show a flatter spectral index in at least two pairs of the frequencies used. No polarised sources in these regions are listed in the López-Caniego (2009) *WMAP* catalogue. On the plane, in the inner Galaxy, the measured spectral indices are consistent with -3.0.

The flatter spectral index in regions No. 1 and 2 between K and Ka bands is interesting as these regions belong to the upper end of the NPS. The flat spectra however, are not observed in β_{K-Q}, which is consistent with -3.0. These two regions have indeed an excess of emission in the Ka band as we can see by the very steep values of β_{Ka-Q}. The low significance of the spectral indices measured with the noisier Ka and Q bands do not allow us to draw strong conclusions about these regions. Nevertheless, the large variation in the spectral index with frequency that we observe in these regions is interesting. Region number 3, which includes the bulk of the emission from the NPS has a steeper spectral index $\beta_{K-Ka} = -3.33 \pm 0.15$ than the one measured in total intensity by Davies et al. (2006) of $-3.07^{+0.09}_{-0.13}$ with an angular resolution of 1°. This might indicate the presence of some AME or free-free emission in the area used in the total intensity analysis.

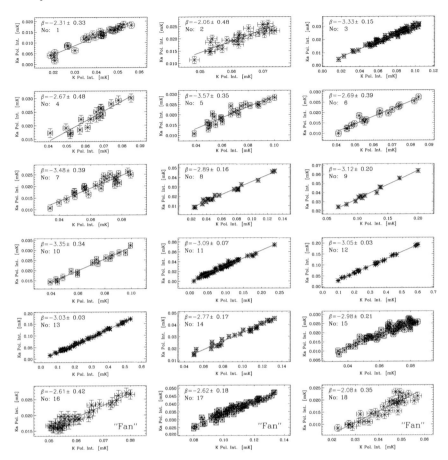

Fig. 3.13 T-T plots of polarisation intensity between 23 and 33 GHz of the eighteen smaller regions defined in Fig. 3.12. The *straight line* represents the best linear fit. The *error bars* show only the statistical fluctuations of each point, the uncertainty in the spectral index β, however, includes the 0.2 % calibration error

Regions 4 and 5, which represent the middle and lower region of the filament next to the NPS show different spectral indices, with region 5 steeper than region 4 in both β_{K-Ka} and β_{K-Q}. Regions 6 and 7, which represent the middle and lower sections of the Galactic centre filament also show a similar behaviour, the lower part of the filament has a β_{K-Ka} steeper than the upper section.

The regions closer to the Galactic plane in the inner Galaxy (regions 8, 9, 10, 11, 12, 13 and 14) have values of β_{K-Ka} consistent with -3.0. These regions represent the area of the sky with the largest amount of polarised emission. A box of 90° in longitude by 20° in latitude, which represents only 4.4 % of the total area of the sky encompasses more than 20 % of the polarised emission at K band. Hence, this

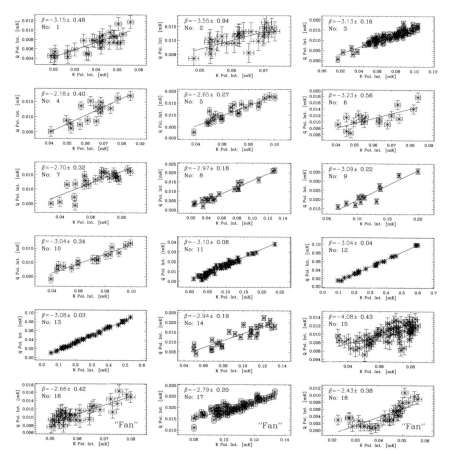

Fig. 3.14 T-T plots of polarisation intensity between 23 and 41 GHz of the eighteen smaller regions defined in Fig. 3.12. The *straight line* represents the best linear fit. The *error bars* show only the statistical fluctuations of each point, the uncertainty in the spectral index β, however, includes the 0.2 % calibration error

region is particularly significant for the estimation of a full-sky average value of the synchrotron spectral index.

Table 3.4 lists the weighted mean of the spectral indices from all the regions. There is a hint of steepening of the spectral index with higher frequencies. The dispersion however, quoted in Table 3.4 is large. There is about the same number of regions that show a steepening in the spectral index as the ones that have a flatter β at higher frequency.

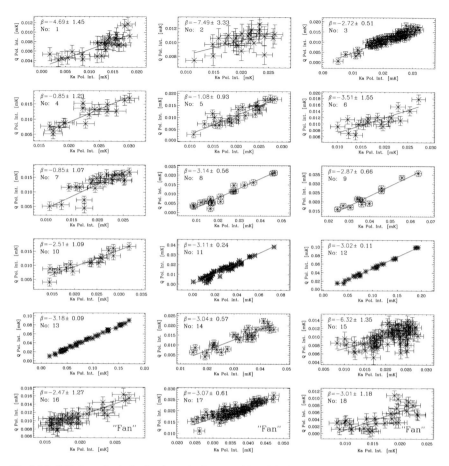

Fig. 3.15 T-T plots of polarisation intensity between 33 and 41 GHz of the eighteen smaller regions defined in Fig. 3.12. The *straight line* represents the best linear fit. The *error bars* show only the statistical fluctuations of each point, the uncertainty in the spectral index β, however, includes the 0.2 % calibration error

3.3.3 Tests on Spectral Indices Values

In order to assess the reliability of the procedure to measure spectral indices, we have tested our method in a set of simulations of the sky. We used the *Planck* sky model (PSM) (see Sect. 2.2) to generate templates of polarised synchrotron of the sky at the *WMAP* frequencies. We generated polarisation maps at K, Ka and Q-bands using three different synchrotron spectral indices, $\beta = -2.5$, $\beta = -3.0$ and $\beta = -3.5$. This spectral index is fixed over the sky. We then added different noise realisations based on the covariance matrix of the *WMAP* data to these polarised synchrotron templates. 500 simulations were generated for each band and different spectral index value.

Table 3.3 Spectral indices in different regions between *WMAP* $K - Ka$, $K - Q$ and $Ka - Q$ bands

Region	l_0	b_0	l_{size}	b_{size}	β_{K-Ka}	χ^2_{red}	q	β_{K-Q}	χ^2_{red}	q	β_{Ka-Q}	χ^2_{red}	q
1	5.0	75.0	30.0	10.0	−2.31, 0.33	0.54	0.97	−3.15, 0.48	0.67	0.89	−4.69, 1.45	0.74	0.82
2	25.0	57.0	10.0	14.0	−2.06, 0.48	1.20	0.23	−3.55, 0.94	0.41	0.99	−7.49, 3.33	0.59	0.94
3	35.0	26.5	10.0	27.0	−3.33, 0.15	0.55	1.00	−3.13, 0.16	0.81	0.87	−2.72, 0.51	0.68	0.98
4	20.0	27.5	10.0	5.0	−2.67, 0.48	1.06	0.39	−2.18, 0.40	2.05	0.01	−0.85, 1.23	0.91	0.55
5	20.0	18.5	10.0	7.0	−3.57, 0.35	1.29	0.17	−2.65, 0.27	1.06	0.38	−1.08, 0.93	0.95	0.52
6	−6.5	21.5	9.0	7.0	−2.69, 0.39	0.48	0.95	−3.23, 0.56	1.80	0.03	−3.51, 1.55	1.38	0.15
7	0.0	13.5	10.0	7.0	−3.48, 0.39	1.09	0.35	−2.70, 0.32	1.94	0.01	−0.85, 1.07	1.34	0.13
8	12.0	7.5	10.0	5.0	−2.89, 0.16	0.90	0.57	−2.97, 0.18	0.70	0.81	−3.14, 0.56	0.99	0.47
9	−6.5	7.5	7.0	5.0	−3.12, 0.20	0.84	0.59	−3.09, 0.22	2.68	0.00	−2.87, 0.66	1.26	0.25
10	−14.5	11.0	7.0	8.0	−3.35, 0.34	0.41	0.97	−3.04, 0.34	0.61	0.86	−2.51, 1.09	0.66	0.82
11*	32.0	0.0	24.0	8.0	−3.09, 0.07	1.95	0.00	−3.10, 0.08	1.42	0.02	−3.11, 0.24	1.49	0.01
12*	10.0	0.0	10.0	6.0	−3.05, 0.03	2.62	0.00	−3.04, 0.04	1.49	0.06	−3.02, 0.11	1.92	0.01
13*	345.0	0.0	20.0	6.0	−3.03, 0.03	2.12	0.00	−3.08, 0.03	1.17	0.20	−3.18, 0.09	0.84	0.76
14	354.0	−11.0	12.0	8.0	−2.77, 0.17	0.83	0.71	−2.94, 0.19	2.74	0.00	−3.04, 0.57	2.07	0.00
15	−2.0	−26.5	16.0	13.0	−2.98, 0.21	0.98	0.51	−4.08, 0.43	1.13	0.22	−6.32, 1.35	1.25	0.09
16	147.5	11.5	15.0	7.0	−2.61, 0.42	0.41	1.00	−2.66, 0.42	0.64	0.93	−2.47, 1.27	0.41	1.00
17*	135.0	0.0	30.0	6.0	−2.62, 0.18	0.87	0.78	−2.79, 0.20	0.87	0.77	−3.07, 0.61	0.82	0.85
18	138.5	−19.0	9.0	12.0	−2.08, 0.35	1.33	0.12	−2.43, 0.38	1.74	0.01	−3.01, 1.18	1.80	0.01

Also listed are the reduced χ^2 of the fit and the χ^2 probability or q-value, q, for each T-T plot

*These regions are on the Galactic plane

Table 3.4 Weighted average spectral indices and the standard deviation of β for the 18 regions

	$\bar{\beta}$	σ_β
$K - Ka$	-3.02 ± 0.03	0.44
$K - Q$	-3.06 ± 0.02	0.41
$Ka - Q$	-3.15 ± 0.07	1.73

We applied our de-biasing procedure to each individual simulation and we calculated the spectral index in the same 18 regions described in the previous section. The idea is to test if we recover the input spectral index.

Figure 3.16 shows the distributions of the recovered spectral indices for the simulations using $\beta_{K-Ka} = -2.5$. All the distributions are centred on the input β values

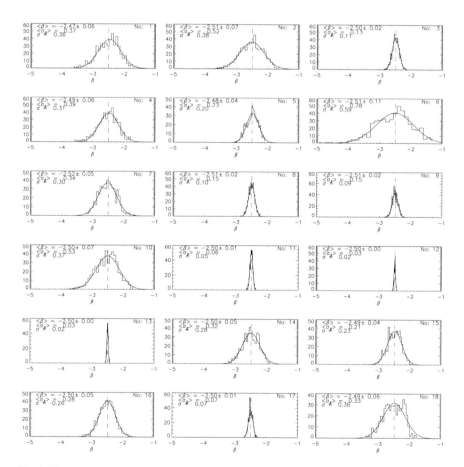

Fig. 3.16 Recovered spectral index distribution between K and Ka bands of the 18 regions defined in Sect. 3.3.2 from 500 simulations. The input value for β_{K-Ka} is -2.5 (shown as a *dashed line*) and is recovered within the uncertainties. Regions with high SNR on the Galactic plane (e.g. No. 11) show a very narrow distribution for β

within one standard deviation. The regions that show the largest discrepancies with
the input value are not necessarily the ones with lower SNR but also the smallest
ones, where the variations due to the small number of pixels is larger. In Fig. 3.17,
histograms for the distribution of the recovered β_{K-Q} for an input value of -3.5 are
shown. This case is more interesting because the steeper value for β makes the SNR
in Q band lower, making this map more susceptible to be affected by noise bias. In
this case only 2 regions shows discrepancies with the input value larger than 1-σ and
all the regions are consistent with the input value at the 2-σ level. We note that in all
the regions, where there is a discrepancy in the spectral index, it is always towards a
flatter value. This might indicate a bias in the estimation of β, but it is a small effect
that only appears at the 1-σ level in the smaller and noisier regions.

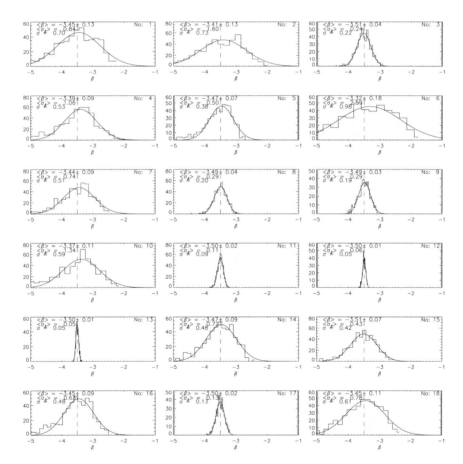

Fig. 3.17 Recovered spectral index distribution between K and Q bands of the 18 regions defined
in Sect. 3.3.2 from 500 simulations. The input value for β_{K-Ka} is -3.5 (shown as a *dashed line*)
and is recovered within the uncertainties

This results shows that our estimation of the spectral index is very robust within the uncertainty. The largest variations are still well between the error bar of the recovered spectral index. A small bias towards flatter values of β might be present is some regions but it is also within the noise fluctuations. In conclusion, we can trust the accuracy of the spectral index measurements that we performed.

3.4 Faraday Rotation

A magnetic field with a component parallel to the line-of-sight will induce the effect of Faraday rotation, in which the plane of polarisation of a travelling photon rotates as it travels through the ISM. The amplitude of the effect depends on the strength of the magnetic field along the line of sight and on the electron density. The observable, the rotation measure RM, is a change in the polarisation angle, χ which scales with the square of the wavelength λ. It is defined as follows,

$$\mathrm{RM} = \frac{\mathrm{d}\chi}{\mathrm{d}\lambda^2}. \tag{3.5}$$

If there is a polarised source in the line-of-sight with no intrinsic Faraday rotation, then RM is equal to the Faraday depth, which is defined as (Burn 1966)

$$\phi = \frac{e^3}{2\pi m_e^2 c^4} \int_0^d n_e(s) B_{||}(s) \, \mathrm{d}s, \tag{3.6}$$

where e and m_e are the electron charge and mass, c the speed of light and the integral is performed over the electron density n_e and the line-of-sight component of the magnetic field $B_{||}$.

Oppermann et al. (2012) constructed a map of the Galactic Faraday depth over the full sky using the most complete data set of Faraday rotation measurements of extragalactic sources to date, including 41330 sources. Their method corrects for the intrinsic Faraday rotation of the extragalactic sources so their final map measures the electron density and radial magnetic field in the Galaxy. Figure 3.18 shows the full-sky Faraday depth map, which has a HEALPIX resolution of $N_{\mathrm{side}} = 128$. The positive values correspond to regions where the radial component of the magnetic field, integrated along the line-of-sight is negative (points to us). We use this map to predict the Faraday rotation at the *WMAP* frequencies and show that is very small. This justifies the assumption regarding a fixed polarisation angle for the synchrotron emission across the *WMAP* bands used in the de-biasing method (Sect. 2.5.2). Using Eq. 3.5 and the Faraday depth map, we can obtain the expected variation in the polarisation angle, $\Delta\chi$, at the different *WMAP* bands. We calculated this number for the entire sky (the map of $\Delta\chi$ has the same morphology as the Faraday depth map in Fig. 3.18, given the linear transformation). In Fig. 3.19 we show the histogram with the distribution of the expected $\Delta\chi$ at K-band. The distribution of pixels is

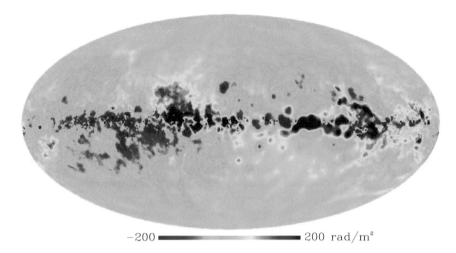

Fig. 3.18 Faraday depth map from Oppermann et al. (2012). We have restricted the linear scale in order to highlight the areas with lower contrast. The total range of the map is $-1224 < \phi < 1453 \, \text{rad} \, \text{m}^{-2}$

Fig. 3.19 Histogram of the expected variation in the χ due to Faraday rotation at K-band over the full-sky. The mean value, its error, and the standard deviation are quoted in the figure

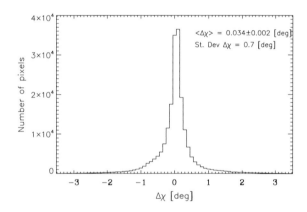

centred at zero and is very narrow, with a standard deviation of $0°.7$. The areas where $|\Delta\chi| > 5°$ correspond to less than 0.22% of the full-sky, and are located on small regions in the Galactic plane with $90° < l < -90°$ i.e. towards the inner Galaxy, where most of the ionised regions of the Galaxy occur. The expected value for $\Delta\chi$ at Ka and Q–band are a factor 0.47 and 0.31 smaller than at K-band respectively.

As the Faraday rotation at *WMAP* frequencies is expected to be less than $1°$ on average, a measurement of the variation in χ between K band and the higher frequencies should indicate only the level of noise in the maps. We measured $\Delta\chi$ between K and Ka over most of the sky using the $1°$ smoothed polarisation maps. We masked out the pixels where the SNR in polarisation is less than 3. The SNR in polarisation is calculated as $P/\sigma_P = \sqrt{Q^2\sigma_Q^2 + U^2\sigma_U^2 + 2QU\sigma_{QU}^2}$. In Fig. 3.20 we show a map of

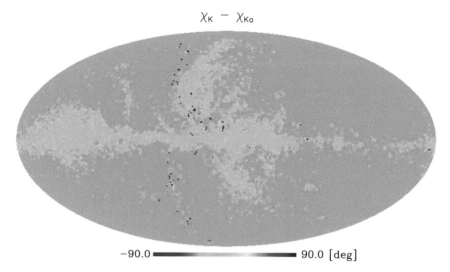

$$\chi_K - \chi_{Ka}$$

-90.0 ■■■■■■■■ ■■■■ $90.0 \ [\text{deg}]$

Fig. 3.20 Map of the difference between the polarisation angle at K and Ka bands. Most of the pixels of the map have a value that is consistent with zero

the difference in polarisation angle between *WMAP*-K and *WMAP*-Ka bands. Most of the pixels show a value very close to zero in this map.

We calculated the weighted average of the polarisation angle for each map in the following way. We first normalise the Q, U vector maps with the debiased polarisation amplitude map P_{deb} to obtain $Q_N = Q/P_{deb}$ and $U_N = U/P_{deb}$. Then, we calculate the weighted average of the Q_{norm} and U_{norm} maps, $\langle Q_N \rangle$ and $\langle U_N \rangle$, using the noise maps over all the valid pixels on the sky (the ones with SNR > 3). Finally, we calculate the averaged polarisation angle using the $\langle Q_N \rangle$ and $\langle U_N \rangle$ values. The calculated averages are $\langle \Delta\chi_{K-Ka} \rangle = -0°7 \pm 0°2$ and $\langle \Delta\chi_{K-Q} \rangle = 0°7 \pm 0°3$. This shows that on average, the Faraday rotation over the entire sky is very small, less than $1°$ at K, Ka and Q bands.

We investigate further by measuring the difference in polarisation angle in smaller areas on the sky. We took the same 18 regions that we used to measure the polarisation spectral indices (shown in Fig. 3.12). We used the same weighted average method but within the pixels inside each box to obtain a single measurement for each region. Table 3.5 lists the location and size of the regions, $\langle \Delta\chi_{K-Ka} \rangle$ and $\langle \Delta\chi_{K-Q} \rangle$ for each region.

The difference in angle between K and Ka bands is consistent with zero in most of the boxes studied. Regions No. 11 and No. 12 are the only ones that show a significant difference in the polarisation angle between K and Ka, here, $\langle \Delta\chi_{K-Ka} \rangle$ is $-6°2 \pm 0°7$ and $-4°3 \pm 1°1$. These regions are interesting because they lie on the Galactic plane, where the SNR is higher. We can discard a Faraday rotation origin for the angle difference in region No. 12, because the observed difference with the higher frequency point, $\langle \Delta\chi_{K-Q} \rangle = -2°5 \pm 1°8$, is consistent with zero,

Table 3.5 Polarisation angle difference averaged in the 18 regions shown in Fig. 3.12, between *WMAP* K – Ka and K – Q bands. $\Delta\langle\chi_{K,Ka}\rangle$ and $\Delta\langle\chi_{K,Ka}\rangle$ are difference of the weighted mean polarisation angle in each region

Region	l_0	b_0	l_{size}	b_{size}	$\Delta\langle\chi_{K,Ka}\rangle$	$\Delta\langle\chi_{K,Q}\rangle$
1	5.0	75.0	30.0	10.0	1.0, 3.1	3.7, 4.7
2	25.0	57.0	10.0	14.0	0.5, 2.0	2.6, 3.2
3	35.0	26.5	10.0	27.0	−2.1, 1.1	−3.0, 1.6
4	20.0	27.5	10.0	5.0	1.9, 2.3	−1.7, 4.1
5	20.0	18.5	10.0	7.0	−0.4, 2.2	−4.2, 3.0
6	353.5	21.5	9.0	7.0	2.2, 2.8	0.8, 3.7
7	0.0	13.5	10.0	7.0	0.9, 2.6	−7.6, 3.6
8	12.0	7.5	10.0	5.0	−0.8, 3.2	1.5, 4.8
9	353.5	7.5	7.0	5.0	0.9, 1.6	1.5, 2.2
10	345.5	11.0	7.0	8.0	0.3, 2.0	−1.5, 3.1
11*	32.0	0.0	24.0	8.0	**−6.2, 0.7**	**−10.1, 1.1**
12*	10.0	0.0	10.0	6.0	**−4.3, 1.1**	−2.5, 1.8
13*	345.0	0.0	20.0	6.0	−1.5, 0.9	1.1, 1.4
14	354.0	−11.0	12.0	8.0	−0.4, 1.2	−1.2, 2.3
15	358.0	− 26.5	16.0	13.0	−0.0, 1.5	2.2, 2.3
16	147.5	11.5	15.0	7.0	−0.4, 3.6	−7.4, 3.5
17*	135.0	0.0	30.0	6.0	−1.5, 2.4	3.2, 2.5
18	138.5	−19.0	9.0	12.0	0.3, 5.6	−7.0, 6.5

*These regions are on the Galactic plane

while a Faraday rotation effect should make this difference larger than $\langle\Delta\chi_{K-Ka}\rangle$. As for region No. 11, Faraday rotation predicts that $\langle\Delta\chi_{K-Ka}\rangle$ should be a factor $(\lambda_Q/\lambda_{Ka})^2 = 0.66$ smaller than $\langle\Delta\chi_{K-Q}\rangle$. Here, $\langle\Delta\chi_{K-Q}\rangle = -10°.1 \pm 1°.1$, so the ratio $\langle\Delta\chi_{K-Ka}\rangle/\langle\Delta\chi_{K-Q}\rangle$ is 0.61 ± 0.16, close to the 0.66 value predicted by Faraday rotation.

Region 11 shows Faraday rotation but its value is not particularly large averaged inside the box. Smaller regions can have larger rotation measure and present Faraday rotation at *WMAP* frequencies. In the next section we study the case of the Galactic centre polarised source.

3.4.1   *Faraday Rotation at the Galactic Centre*

One region that shows a larger variation in angle corresponds to the Galactic centre. It can be seen as a yellow spot at the centre of Fig. 3.20. Here, $\Delta\chi_{K-Ka} = 21° \pm 3°$, which is calculated over the 4 central pixels of the map (the pixels closest to $l, b = [0°, 0°]$). Here, we measure the polarisation angle of the central region in all the *WMAP* frequencies (at a common 1° resolution) as the large SNR of the source

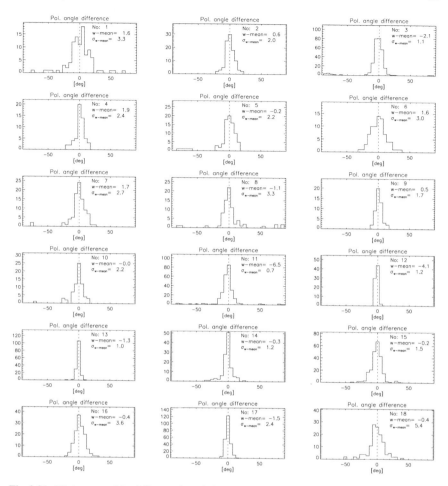

Fig. 3.21 Histograms of the difference in polarisation angle between *WMAP* K-band and Ka–band. The different regions are the ones shown in Fig. 3.12

allows us to do this. We measure the *Q*, *U* Stokes parameters in a circular aperture of 1° in diameter, centred at the location of the source (Figs. 3.21 and 3.22).

In Table 3.6 we list the measurements for all the *WMAP* bands. In Fig. 3.23 we plot the polarisation angle as a function of the square of the wavelength, λ^2. We calculated the RM value, using a least square fit to the angle measurements. The fitted line, with a slope that corresponds to RM$= -4382 \pm 204 \,\text{rad}\,\text{m}^{-2}$ is shown as a dashed line. The angle measured at W-band shows the largest deviation from the line. The RM value that we found is comparable with a measurement by Tsuboi et al. (1995) of the same region, but a much smaller beam size of 39″. They found RM $= -3120 \pm 188 \,\text{rad}\,\text{m}^{-2}$ between 10 and 42.5 GHz. The difference between this value and this result may be related to the larger *WMAP* beam size (effectively 1° in this case), which includes regions with different intrinsic polarisation angles.

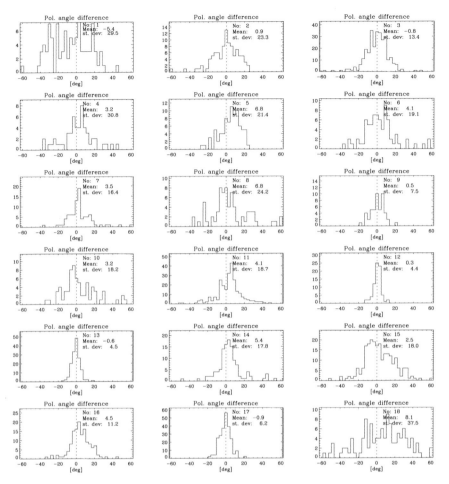

Fig. 3.22 Histograms of the difference in polarisation angle between *WMAP* K-band and Q–band. The different regions are the ones shown in Fig. 3.12

3.5   Polarisation Fraction

Calculating the polarisation fraction, Π, of synchrotron emission is a good way to measure the degree of order of the magnetic field perpendicular to the line-of-sight. This calculation is not trivial because of the difficulty in obtaining the synchrotron total intensity at GHz frequencies. Free-free and AME contribute to the total emission so component separation methods are necessary. Here we compare the polarisation fraction at K-band using different templates for the synchrotron amplitude over the full sky.

Table 3.6 Measured polarisation parameters for the Galactic centre source at the five *WMAP* frequencies

Band	Stokes Q (mK)	Stokes U (mK)	Pol. Angle (deg)
K	-0.43 ± 0.01	-0.91 ± 0.06	57.7 ± 1.6
Ka	-0.57 ± 0.01	-0.25 ± 0.03	78.2 ± 0.8
Q	-0.50 ± 0.01	-0.08 ± 0.01	85.6 ± 0.5
V	-0.27 ± 0.01	0.04 ± 0.01	94.4 ± 1.0
W	-0.05 ± 0.01	0.04 ± 0.01	109.7 ± 4.2

The table lists the Stokes Q and U parameters and the derived polarisation angle. The values correspond to the median value of a $1°$ in diameter aperture centred at $(l, b) = [0°15, -0°1]$. The uncertainties in Q and U correspond to the statistical fluctuations in the map, measured in a larger aperture around the source

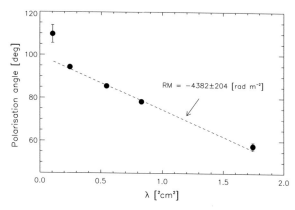

Fig. 3.23 Relation between the polarisation angle as a function of the wavelength squared for the Galactic centre source. The *black dots* correspond to the measured polarisation angle at the five *WMAP* bands. The *dash-line* is a linear fit to the points, expected for a pure foreground Faraday screen, where the polarisation angle $\chi \propto RM \lambda^2$. The derived rotation measure is RM$= -4382 \pm 204$ rad m^{-2}

We also calculate Π in a small region on the Galactic plane where we can precisely calculate the free-free component by using hydrogen radio recombination lines, so a clean synchrotron temperature map can be obtained.

3.5.1 Full Sky Polarisation Fraction

To obtain the polarisation fraction, we measured the ratio between our debiased polarisation map at K-band (produced using the Wardle & Kronberg estimator from Eq. 2.22) and a template for the synchrotron amplitude at this frequency,

$$\Pi = \frac{P_K}{I_K} \tag{3.7}$$

As a first step, we used the total intensity map at K-band only. We subtracted the contribution from the CMB using the internal linear combination (ILC) map provided by the *WMAP* team (Bennett et al. 2013). The ILC map is produced using a weighted combination of the five *WMAP* intensity maps, all smoothed to an angular resolution of 1°. The weights are calculated by minimising the variance of the of the measured temperatures, and the sum of the weights has to be equal to 1. The synchrotron polarisation fraction obtained this way is only reliable for the diffuse emission at high Galactic latitudes, where the free-free and AME contributions are assumed to be small compared with the synchrotron. At K-band, synchrotron dominates over 75 % of the sky and in regions like the NPS, it is at least 15 times brighter than free-free and dust emission. Figure 3.24 shows a map of the polarisation fraction with an angular resolution of 3°. Π is high ($\sim$40 %) along the NPS and other filaments, even after the significant smoothing. This shows that the magnetic field on these structures is highly ordered. The uncertainty of this map is estimated using the uncertainties maps for temperature and polarisation maps. It is dominated by the uncertainty in the polarisation amplitude map. The median value over the sky is 9.5 %.

A first way of estimating the synchrotron intensity is by extrapolating the 408 MHz map up to 23 GHz. At 408 MHz there is some free-free coming mainly from star forming regions on the Galactic plane. We subtracted this component by scaling the

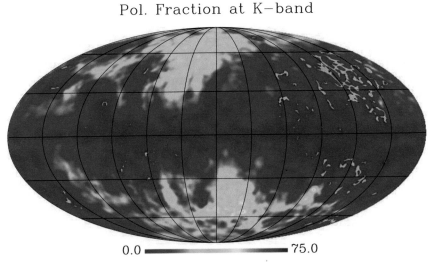

Pol. Fraction at K−band

0.0 ▬▬▬▬▬▬▬ 75.0

Fig. 3.24 Synchrotron polarisation fraction map at K-band. We subtracted the ILC map of the CMB fluctuations from the total intensity map. Free-free and AME have not been subtracted from the total intensity map. Therefore, this map for synchrotron polarisation fraction is only reliable in regions where synchrotron dominates such as the NPS. The colour-scale ranges from 0 up to 75 %. The median uncertainty in this map is 9.5 %

WMAP MEM map free-free template to 408 MHz using a free-free spectral index $\beta_{ff} = -2.13$, valid for the diffuse gas (Davies et al. 2006). Then, we scaled the resulting pure synchrotron map from 408 MHz to 23 GHz, using a fixed synchrotron spectral index of $\beta_s = -3.0$ (as measured by Davies et al. 2006). The largest source of uncertainty comes from the fixed synchrotron spectral index between 0.408 to 23 GHz. A variation in the synchrotron spectral from $\beta_s = -3.1$, to $\beta_s = -2.9$ decreases the measured polarisation fraction by a factor $\sim$2.

A second approach to obtain the synchrotron polarisation fraction is based on the modelling of the different emission mechanisms in total intensity. Closer to the plane, there is a large free-free contribution from HII regions so it is necessary to subtract it in order to obtain the synchrotron emission. The *WMAP* team provides all-sky foreground templates at each of the five *WMAP* frequencies, generated using two procedures: Maximum Entropy Method (MEM) and Monte Carlo Markov Chains (MCMC) fits (for a detailed description of the methods, see Gold et al. 2011). The MEM technique generates synchrotron, free-free, spinning dust, and thermal dust templates and assumes that the spectral indices of these foregrounds are constant over the sky. The MCMC method produces four different synchrotron templates, one for each sets of parameters of the prior model. (See Bennett et al. 2013, for a description of the models investigated.) All the methods used to calculate the synchrotron total intensity are listed in Table 3.7.

We masked three regions on the synchrotron templates that are problematic for calculating the polarisation fraction. First, Tau A ($l = 184°.3$, $b = -5°.8$), where the synchrotron spectral index is flatter ($\beta_s \approx -2.3$) than that observed in the diffuse gas; the ζ Oph HII region ($l = 6°.3$, $b = -23°.6$), where the synchrotron contribution is clearly underestimated in some MCMC models methods; and an area from the Gum nebula ($l = 254°.5$, $b = -17°.0$), where the same underestimation of synchrotron occurs.

Figure 3.25 shows the six different polarisation fraction maps. The differences in these maps are due only to the different synchrotron intensity template used. The extrapolation of the free-free corrected Haslam et al. (1982) map (top-left) produces a Π map which is morphologically very similar to the polarisation amplitude

Table 3.7 Methods used to estimate the synchrotron total intensity at 23 GHz

Method	Note	σ_Π (%)
$(408\,MHz - ff) \rightarrow 23\,GHz$	Assumes fixed $\beta_{syn} - 3.0$	14
WMAP MEM	Includes AME. Fixed $\beta_{syn} = -3.0$	–
WMAP MCMC model c	No AME component. β_{syn} can vary	37
WMAP MCMC model e	Includes AME. Fixed $\beta_{syn} = -3.0$	21
WMAP MCMC model f	Includes AME. β_{syn} can vary	33
WMAP MCMC model g	Includes AME. β_{syn} varies as described in Strong et al. (2011)	33

Also listed is the median uncertainty over the sky of the polarisation fraction derived form each method. The MEM maps does not have a quantification of its uncertainty

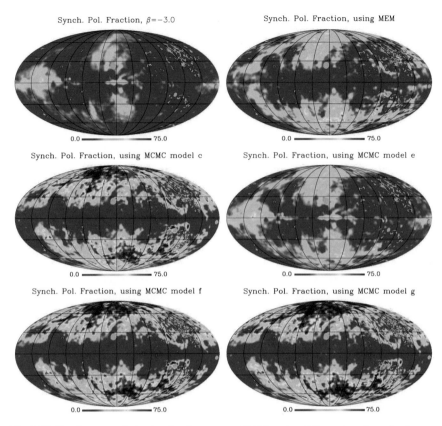

Fig. 3.25 Synchrotron polarisation fraction maps at 23 GHz using different templates for the synchrotron intensity (see text). The colour scale is linear, and it ranges from 0–75 %. The preferred Π map is thought to be MCMC model-e

map, where the structures are filamentary; regions with high polarisation fraction are individual filaments or specific areas on the plane (e.g. the fan region at $l \sim 140°$). The one created using the free-free MEM model (top-right) has higher polarisation fraction and also shows a bigger difference between mid and low Galactic latitudes, where the polarisation fraction is near to zero on the plane. The output with the MCMC model e is very similar to the one using the MEM model at high latitudes and to the Haslam map extrapolated close to the plane. It was created with a fixed synchrotron spectral index of -3.0. MCMC models c, f and g produce very similar and incorrect values for Π, as they result in polarisation fractions $\sim 100\%$ over large areas of the sky at high Galactic latitudes. We therefore assume that these maps are biased over large areas of the sky and we will not discuss them anymore.

Some of the filaments are highly polarised ($\Pi \approx 40\%$) regardless of the synchrotron model used. We believe that qualitatively, the best model for the synchrotron total intensity is the MCMC model-e (on the right in the middle row of Fig. 3.25).

This is because the polarisation fraction at high latitudes is very similar to the one from Fig. 3.24. The map in Fig. 3.24 represents an upper limit for the synchrotron polarisation fraction. Therefore we discard the maps that predict much larger values at high latitudes (models c, g, g). Moreover, on the plane, model-e is similar to the prediction using the Haslam map with $\beta = -3.0$, value that is similar to the average spectral index of synchrotron that we measured on Sect. 3.3.

In general, the uncertainty in these full sky polarisation fraction maps is large ($\sim$20 %) due to uncertainties in the free-free contribution and the synchrotron spectral index. In the next Section, we investigate the Galactic plane region around $l = 140°$ in more detail. Here, a detailed free-free map is available which can be used to estimate the true synchrotron emission.

3.5.2 Polarisation Fraction in the $20°\!.0 \leq l \leq 44°\!.0$ Range

As we discussed earlier, the main problem in the calculation of the synchrotron polarisation fraction at *WMAP* frequencies is the determination of the true synchrotron total intensity. The mix of different components present at these frequencies makes it difficult to isolate the synchrotron component. On the Galactic plane, the problem is more acute due to the superposition of different regions (e.g. HI gas, HII regions, molecular clouds, SNR, AME clouds) along the line-of-sight.

At K-band, the main emission mechanisms in total intensity on the Galactic plane are AME, synchrotron and free-free. The free-free contribution can be estimated at high Galactic latitudes using Hα maps, but on the plane, dust absorption make this prohibitive. Alves et al. (2012) produced a free-free template on the Galactic plane using hydrogen radio recombination lines (RRL) for the region inside $20° \leq l \leq 44°$, $-4° \leq b \leq 4°$. This map has an angular resolution of $14'\!.8$ and a calibration uncertainty of $\approx$10 % for extended sources.

A synchrotron total intensity map at K-band can be obtained in the following way using the RRL free-free template. First, we scaled the free-free map to 408 MHz using a free-free spectral index $\beta_{ff} = -2.13$ (Davies et al. 2006). Then, we subtract the free-free contribution at 408 MHz. Finally, we scale this pure synchrotron map up to 23 GHz using a synchrotron spectral index $\beta_{syn} = -3.0$. An additional map of AME can be obtained by subtracting the synchrotron and free-free components from the total intensity map. At 23 GHz on the Galactic plane, the CMB anisotropy contribution is small ($\Delta T \approx 0.5$ mK) so it can be ignored. In Fig. 3.26 we show the different maps produced for this region. On top are the *WMAP* temperature and polarisation maps at K-band. A bright source can be seen in the polarisation map, at $(l, b) = [34°\!.6, -0°\!.5]$, the SNR W44 associated with the pulsar PSR B1853+01. This source has a polarisation fraction of 17 ± 4 % and a spectral index of -2.12 ± 0.04 (Frail et al. 1996). In the middle panel of Fig. 3.26, are shown the 408 MHz map scaled to K-band using $\beta_{syn} = -3.0$ and the RRL free-free map scaled to K-band from Alves et al. (2012). The bottom panel show the pure synchrotron and AME maps that we produced.

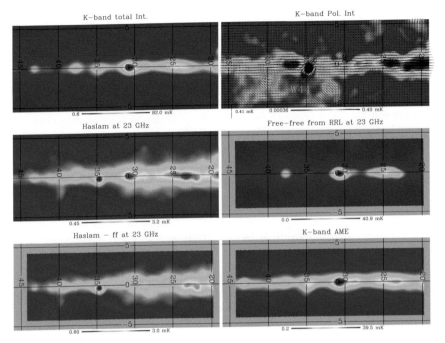

Fig. 3.26 Maps of the Galactic plane within $20° \leq l \leq 44°$, $-4° \leq b \leq 4°$. On *top* are the *WMAP* total intensity and polarisation intensity maps at K-band. The magnetic field vectors are also plotted in the *right* and the position of the W44 SNR is indicated. In the *centre*, on the *left*, is the 408 MHz map scaled to 23 GHz using $\beta_{syn} = -3.0$. On the *right* is the free-free map based on RRL from Alves et al. (2012). At the *bottom-left* is the free-free-corrected synchrotron template obtained by subtracting the free-free from the Haslam et al. (1982) map. At the *right* is the AME template produced by subtracting the free-free and synchrotron components from the total intensity template. All maps have an angular resolution of $1°$ and the grid spacing is $5°$

To calculate the polarisation fraction in the $20° \leq l \leq 44°$ region, we averaged pixels with equal latitude to create cuts across the Galactic plane using the different maps shown in Fig. 3.26. Because the spectral index of the W40 SNR visible in the *WMAP* polarisation map is different from the one of the diffuse synchrotron emission on the plane, we masked-out a strip of $3°$ width in longitude centred at the pulsar location.

In Fig. 3.27 are shown the normalised averaged cuts for the total intensity, the polarisation intensity, the 408 MHz Haslam map, the free-free map from the RRLs, the free-free corrected synchrotron and the AME maps. We have modelled these profiles using a parabolic and a Gaussian curve. The Gaussian component accounts for the narrower and central region of the cut while the parabola models the broader component. The best fits are shown as dotted lines for each component in Fig. 3.27. In the synchrotron temperature profile (blue), the broad and narrow component (within $|b| \lesssim 1°$) are evident. The polarised synchrotron cut (red) on the other hand, seems to lack the broader component. However, the Gaussian components of the synchrotron

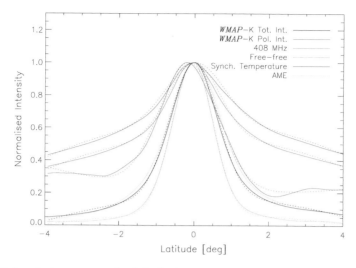

Fig. 3.27 Longitude cuts across the Galactic plane, averaged between $20° \leq l \leq 44°$ and normalised to one. In *black* is the total intensity profile and in *red* is the polarisation amplitude, both at 23 GHz. In magenta is the profile of the 408 MHz Haslam et al. (1982) map. In *green* is the profile of the RRL free-free template. In *blue* is the 408 MHz map corrected by free-free and in *light blue* is the AME component, obtained by subtracting the free-free and synchrotron emission from the total intensity map. The *dashed lines* show the fit to each component, which is the sum of a parabolic and a Gaussian function. The width of the narrow (Gaussian) component for each fit are listed in Table 3.8

Table 3.8 FWHM in degrees of the narrow component of the longitude averaged profiles shown in Fig. 3.27

Map	FWHM	Beam-deconvolved FWHM
WMAP –K Tot. Int.	1.5 ± 0.1	1.1 ± 0.1
WMAP –K Pol. Int.	1.8 ± 0.1	1.5 ± 0.1
408 MHz	1.6 ± 0.1	1.2 ± 0.1
Free-free	1.3 ± 0.1	0.8 ± 0.2
Synch. Temperature	1.8 ± 0.1	1.5 ± 0.1
AME	1.6 ± 0.1	1.3 ± 0.1

All these maps haven a common angular resolution of $1°$ and we also include the beam-deconvolved width for each profile

temperature and polarisation show the same width of $1°\!.8 \pm 0°\!.1$ FWHM. In Table 3.8 we list the FWHM of the narrow components for all the cuts. We also include a beam-deconvolved width for each component.[1] There, we can see that the narrower component of the synchrotron emission is thicker than the narrower component of AME and free-free.

---

[1]This is calculated using $\text{FWHM}_{\text{deconv}} = \sqrt{\text{FWHM}_{\text{obs}} - 1°}$, where FWHM_{obs} is the measured width of the profile. The angular resolution of the maps is $1°$.

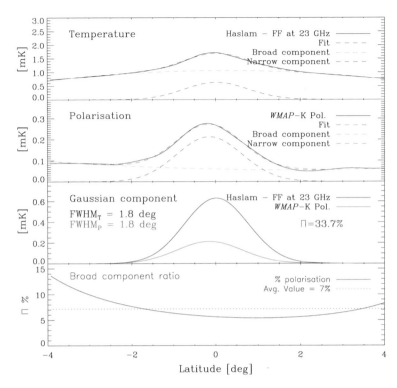

Fig. 3.28 Latitude cuts averaged between $20° \leq l \leq 44°$. The *top panel* shows the synchrotron temperature template, i.e. the Haslam map corrected by free-free and scaled to K-band. In *red* is a fit consisting on a Gaussian plus a parabolic curve. In *green* and *blue* are the two components of the fit. The *second panel* shows the polarisation amplitude cut at K-band on *black*. The fit to the data is in *red* and the *green* and *blue lines* show the fit components. On the *third panel* are the Gaussian components of the previous fits. The polarisation fraction of these narrow components has a peak of 33.7 %. On the *fourth panel* is plotted the polarisation fraction of the broad component which has a mean value of 7 %

On Fig. 3.28 are shown the synchrotron profiles, both temperature and polarisation and the comparison of the narrow and broad components of the fits. The profiles are well described by the parabola plus Gaussian fit. The third and fourth panels on Fig. 3.28 shows the narrow (Gaussian) and broad (parabolic) components. The ratio between the narrow components of the synchrotron polarisation and total intensity has a peak of $\Pi = 33.7\,\%$. The broader component has an average polarisation fraction of 7 %. The 10 % calibration uncertainty on the free-free map induces an error of $\approx 3\,\%$ on the polarisation fraction. The dominant source of uncertainty comes from the synchrotron spectral index between 408 MHz and 23 GHz. A deviation on the spectral index of $\Delta\beta = \pm 0.1$ i.e. $\beta = -3.0 \pm 0.1$, translates into an error in the polarisation fraction of 12 %.

We also calculate the polarisation fraction on the plane without the separation in a narrow and broad components. We use our free-free corrected synchrotron template as well as the *WMAP* MEM and MCMC-e synchrotron templates (MCMC-e produces the best synchrotron polarisation fraction map over the entire sky as we shown in Sect. 3.5.1). We show the latitude cuts in the $l = [20°, 44°]$ range of these three synchrotron templates in the top panel of Fig. 3.29. There is a large difference between the three models. The MEM prediction is much larger (up to a factor $\sim$10 at $b = 0$) than the other two models. The free-free corrected 408 MHz map is very similar to the MCMC-e model but they differ in the central region with $|b| < 1°$, where the MCMC-e is flat and the 408 MHz presents a narrow component.

In the bottom panel of Fig. 3.29, we show the polarisation fraction profiles using the previous three curves. The MEM model predicts a very low polarisation fraction. Kogut et al. (2007) used a similar MEM model to estimate the polarisation fraction on the Galactic plane 2–4 %

Π has a peak value of $\Pi = 16.5\,\%$ on the plane and, in this longitude range, this maximum is centred at $l \approx -1°$. Π decreases when moving away from the plane, having a minimum around $|b| = 2°$. At higher latitudes, Π increases again. The black line of the figure show the polarisation fraction, but this time using the *WMAP* MEM synchrotron template.

This can be seen in the top panel of Fig. 3.29, where the *WMAP* MEM predicts a much larger overall synchrotron component than the extrapolation of the free-free corrected 408 map. However, the narrow polarisation component at $|b| \lesssim 1°$ is not accounted for by the MEM template and the polarisation fraction has a qualitative different shape. There is an excess of $\approx$8 % in Π inside the narrow $|b| \lesssim 1°$ band over the average in $|b| < 4°$.

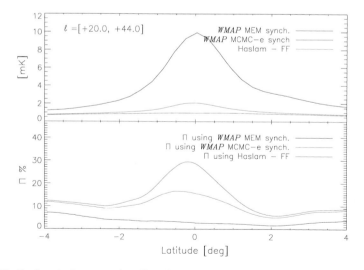

Fig. 3.29 *Top* Longitude averaged profiles of two templates for the synchrotron total intensity. In *black* is shown a cut of the *WMAP* MEM synchrotron map and in *red* is the extrapolation of the free-free corrected 408 MHz to 23 GHz. *Bottom* Polarisation fraction profiles at 23 GHz calculated as the ratio between the *WMAP* polarisation amplitude profile and the two synchrotron total intensity templates shown above

The high polarisation fraction measured on the narrow component (1°.8 thick) on the Galactic plane suggests a very ordered magnetic field that runs parallel to the plane. The regions where the polarisation fraction is low, around $|b| = 2°$, can be explained by the superposition of features with different polarisation angles than that of the ordered component of the plane (e.g. filaments and spurs visible in Fig. 3.32). This depolarisation effect is similar to what occurs around a strong polarised source on the Galactic plane when its polarisation angle is inclined with respect to the plane emission.

We also produced cuts parallel to the Galactic plane, shown in Fig. 3.30. The plots show the longitude cuts below the plane, at $b = -2°.3$. We can see that the polarisation fraction on angular scales of $1°$ has large variations, from $\sim 0\,\%$ up to $20\,\%$. The region with the smallest polarisation fraction ($25° \gtrsim l \gtrsim 30°$) has some filaments running perpendicular to the Galactic plane where the line-of-sight depolarisation occurs. The profile of these filaments is visible in total intensity in the middle panel, on top of the smooth synchrotron background, but they are not present in polarisation. Figure 3.31 shows the polarisation amplitude at 23 GHz in the $l = [20°, 44°]$, $b = [-15°, 15°]$

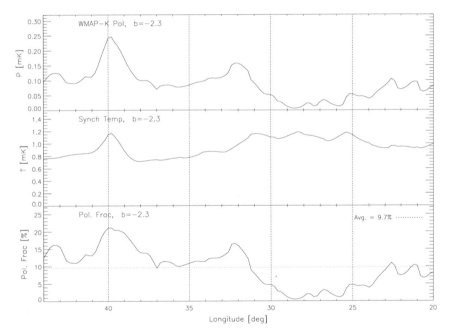

Fig. 3.30 Longitude cut at $b = -2°.3$ of synchrotron polarisation, total intensity from the free-free corrected 408 MHz data and polarisation fraction at K-band. Notice the low polarisation between $25° \gtrsim l \gtrsim 30°$. This is a depolarisation strip resulting from the superposition of emission from filaments with polarisation angle perpendicular to the emission parallel to the Galactic plane (see Fig. 3.31). On the *middle panel*, the filaments can be seen above the smoother synchrotron background from the plane

Fig. 3.31 Polarisation amplitude at K-band on a region that includes the RRL mapped area $l = [20°, 44°]$, $b = [-4°, 4°]$. Notice the depolarisation zones around $b = +2°$ and $b = -2°$ due to the superposition of emission with different polarisation angles, in this case the Galactic plane with some roughly perpendicular filaments

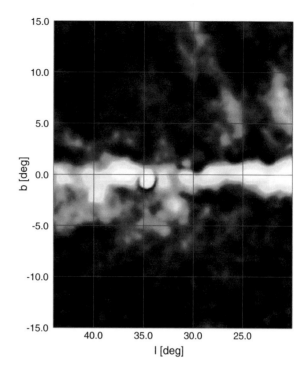

range. The depolarisation region at $b \approx 2°$ and $b \approx -2°$ around $l \sim 25°$ occurs in the intersection of some large filaments with the Galactic plane.

The previous analysis shows that the polarisation fraction of the diffuse synchrotron at 23 GHz can be as high as 25 % on some regions on the Galactic plane on 1° angular scales. Kogut et al. (2007) quote a synchrotron polarisation fraction of 2–4 %, using the *WMAP* MEM model for the synchrotron total intensity at an angular resolution of 7° on the plane. The larger smoothing of the maps in their analysis smears out some polarisation signal by mixing regions with different polarisation angle on the beam. We note however than without the extra smoothing, there are significant differences in the polarisation fraction on the plane because of the difference between the MEM synchrotron model and the free-free corrected extrapolation of the 408 map.

A remark to be made is that the variations in the polarisation fraction in this region of the plane will appear naturally following the method used here if the spectral index of the emission is not constant. We have shown in the Sect. 3.3 that the spectral index of the synchrotron polarisation can show variations on 3° scales, so we expect that on smaller scales these variations could be more pronounced. This shows that the diffuse synchrotron from the Galactic plane has more variations than previously claimed (e.g. Kogut et al. 2007).

3.6   Discussion

During this chapter we have presented some observational parameters that we can derive from the *WMAP* polarisation data. We have seen that an important part of the polarised emission at microwaves wavelengths comes from filamentary structures at high Galactic latitudes. What is the origin of these filaments? How do they relate to the rest of the ISM, the neutral gas for example? We start this discussion by looking at HI data, which traces the neutral gas of the Galaxy. We then go through the hypothesis for the origin of these filaments and use *WMAP* data to test the predictions of one model.

We also include a section in which we estimate the contribution of the polarised emission from the filaments in the context of foreground to the CMB.

3.6.1   HI Morphology

Most of the filamentary features that extend roughly perpendicular to the Galactic plane lie in the inner Galaxy, with $|l| \lesssim 60°$. In these structures, the polarisation angle is well aligned along the extension of the filaments. The top panel of Fig. 3.32 shows the inner region of the Galaxy between $|b| < 30°$ visible in K-band polarisation.

These structures appear similar to the HI "worms" described by Heiles (1984), Koo et al. (1992). Heiles proposed that these worms observed in the inner Galaxy are shells with open tops. Stellar winds and supernova explosions create shells in the ISM and if the expansion energy is large enough, hot gas will flow out through chimneys to the Galactic halo. Chimney walls are vertically supported by magnetic fields (Breitschwerdt et al. 1991), which can glow in radio continuum if cosmic rays diffuse through them.

Sofue (1988) describes a number of radio spurs present in the 408 MHz map of Haslam et al. (1982), and most of these spurs lie in the inner Galaxy. He did not find definitive correlations between HI and the radio continuum structures in this region. Here, we examine the more sensitive HI 21 cm maps from the Leiden/Argentine/Bonn (LAB) survey (Kalberla et al. 2005), which maps the full-sky with an angular resolution of 36 arc min FWHM in the velocity range from -450 to $+400$ km s^{-1}, at a resolution of 1.3 km s^{-1}. We found no obvious morphological correlations between the polarisation spurs and the HI velocity maps. The only features that have HI counterparts are the NPS and the polarised filament No. 10 in our nomenclature. Figure 3.32 *bottom* shows the polarisation map with the HI contours of the velocity range that shows the correlation. From the figure, we can see that the correlation is better near the base of the filament and with increasing latitude, there is a spatial separation in the HI emission from the polarised synchrotron. This HI filament in particular is part of a larger shell, probably the limb-brightened periphery of a HI bubble. This is visible in one of the panels of Fig. 3.33 where we show the full sky

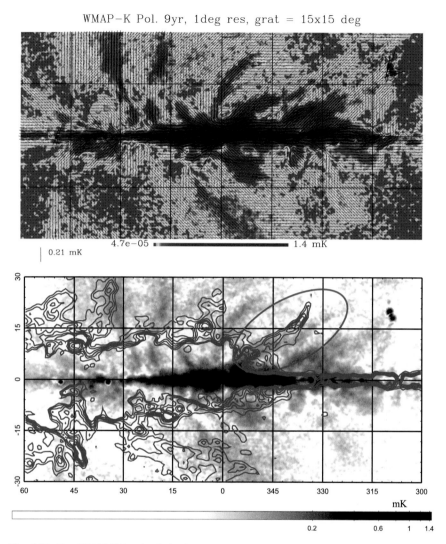

Fig. 3.32 *Top WMAP* K-band polarisation intensity map showing the inner Galaxy between $l = \pm 60°$. The polarisation vectors are rotated by 90° to indicate the direction of the magnetic field. *Bottom* Same as the map on *top* but using a linear colour scale. The contours of HI 21 cm emission at $v = +10$ km s^{-1} are shown in magenta. There is a possible association between the HI and the polarised intensity on the filament that starts close to the Galactic centre and runs towards negative longitudes above the Galactic plane. The angular resolution is 1° and the graticule has 15° spacing in both Galactic coordinates

HI maps that have velocities close to zero. In the panel with $v = +10$ km s^{-1} there is a shell-like structure of ~30° in diameter centred around $(l, b) = (350°, 20°)$.

If the polarised emission that comes from Filament No. 10 is indeed physically correlated with the HI emission from this bubble, this would resemble the Heiles

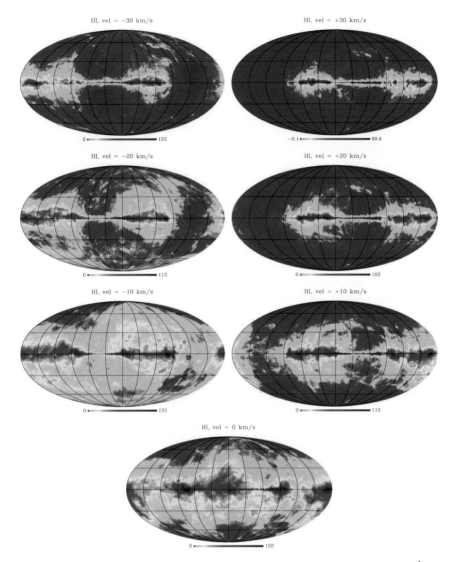

Fig. 3.33 HI maps of the full sky at different line-of-sight velocities in the range $\pm 30\,\mathrm{km\,s^{-1}}$. The graticule has $30°$ spacing in both Galactic coordinates and the maps are plotted using a histogram-equalised scale

(1984) picture previously described. Here, an enhancement in the magnetic field, produced by the compression of the ambient field by an expanding shock, can induce the emission of polarised synchrotron radiation along the field lines.

On larger angular scales, we can see what happens on the full-sky. In Fig. 3.33 we have plotted the HI maps for seven different velocity ranges, between $\pm 30\,\mathrm{km\,s^{-1}}$.

At velocities of -10 and 0 km s^{-1}, a filament coincident with the NPS can be seen at $l = 30°$. This neutral gas counterpart of the NPS has been known for decades (see e.g. Heiles et al. 1980).

We draw attention to the structure centred at $l \sim 320°$ on the plane, the large shell that is most visible at -10 km s^{-1}. This expanding shell is know as the Scorpius-Centaurus super-shell, a cavity in the local ISM which is thought to be the expanding remnants of a number of SNe explosions (e.g. Weaver 1979). Inside this cavity lies the Sco-Cen OB association, the nearest to the Sun group containing massive stars, where the distances of its members lie between 118 and 145 pc and the oldest groups of star have an age ~ 15 Myr (Preibisch and Mamajek 2008). We note that the HI counterpart of the NPS lies on the periphery of the Sco-Cen shell. We will discuss this connection in more detail in the next section.

3.6.2 The Origin of the Large Scale Polarised Loops

Since the early attempts to explain the origin of the radio continuum loops, SN remnants expanding in the Galactic magnetic field have been a preferred alternative. Spoelstra (1971, 1972, 1973) applies the model of a sphere expanding in the ISM by Laan (1962) to Loops I, II, III and IV. In this model, the compression of the magnetic field due to the shock wave from the SN will act as a synchrotron source. The emission will be more easily observable on the periphery of the shell due to limb-brightening, which explains the loop-type shape of the remnants.

Sofue et al. (1974) discuss a number of difficulties that this SNR hypothesis faces to explain diverse observational data. They present a new model in which the spurs are the result of the tangential view of a region of shocked gas produced at the spiral arms that extend above and along the arms. This hypothesis explain the origin of Loops II and III and many other spurs according to the authors. The NPS on the other hand, can be explained according to Sofue (1977) in terms of a magnetohydrodynamical phenomenon associated with the Galactic centre. A similar scenario connected to the Galactic centre was more recently proposed by Bland-Hawthorn and Cohen (2003), based on diffuse filamentary infra-red emission observed on both sides of the Galactic plane, which is thought to be formed by a central outburst with an energy scale $\sim 10^{55}$ ergs. This interpretation implies that the spur has a kpc scale, in comparison with the hundred-pc scale that is expected from the SNR hypothesis.

It is clear that a distance determination for the spurs was necessary to settle which interpretation was correct. Bingham (1967), using optical polarisation data, which is correlated with the polarised emission from the NPS, determined a distance of 100 ± 20 pc. The discovery of HI emission from the periphery of the NPS in the maps of Heiles (1974) and Colomb et al. (1980) has also helped in this matter. Puspitarini and Lallement (2012) have shown that HI gas that belongs to the top region ($b \geq +70°$) of Loop I is located at a distance of 98 ± 6 pc and the material at intermediate latitudes ($+55° \leq b \leq +70°$) is around 95–157 pc. This is strong evidence that favours a nearby shell as the origin of Loop I.

In this nearby scenario, a problem with a single SN event as the origin for Loop I is the discrepancy on the determination of the age for such SN event. The low expanding velocity of the HI gas, $|v| \leq 20 \, \text{km s}^{-1}$ (Kalberla et al. 2005; Sofue et al. 1974; Weaver 1979), implies an SN older than several 10^6 yr. On the other hand, soft X-ray emission detected from the interior of the radio loop by Bunner et al. (1972) suggest an age 10 times younger. Moreover, the initial SN energy in either case is $\sim 10^{52}$ ergs, an order of magnitude larger than the standard 10^{51} ergs (Egger and Aschenbach 1995).

A related scenario in which instead of a single SN event, the Loop I cavity is a super-bubble, the result of stellar winds and consecutive supernovae in the Sco-Cen OB association is more attractive. This idea has been suggested by a number of authors (see e.g. Heiles et al. 1980; Weaver et al. 1977). Egger (1995) present a model in which a recent SN inside the Sco-Cen super-shell shocks the inner walls of the bubble, giving rise to the NPS emission.

A expanding super-shell in a magnetised medium has been modelled by Tomisaka (1992). In their model, during the early stages of the shell (a few Myr), the expansion is spherically symmetric. In a later phase (~ 50 Myr), the magnetic field stops the expansion perpendicular to the field lines producing a highly elongated bubble. Cosmic rays accelerated by the shock will emit polarised synchrotron travelling through the distorted field lines produced by the shell.

Assuming that the unperturbed magnetic field lines in the vicinity of the Sun are parallel to the Galactic plane, an expanding super-shell will bend the lines in a simple manner. The originally parallel field lines will follow lines of constant longitude on the surface of the expanding sphere. The observed pattern from Earth however is not trivial and it will depend on the viewing angle of the shell. Following what Heiles (1998) showed using starlight polarisation, we can compare the direction of the field lines in this scenario, produced by a shell centred at the location of the Loop I bubble, with the polarisation angle of the emission seen by *WMAP*. This type of modelling was also used by (Wolleben 2007), who placed two overlapping shells that deform the magnetic field medium to reproduce the polarisation directions observed at 1.4 GHz. These data show a depolarisation band confined at $|b| \approx 30°$, so their fit is based only on the high latitude data. The low Faraday rotation of the *WMAP* data allow us to compare directly the angles in the $|b| \leq 30°$ band, where there is strong emission from the spurs.

We locate the centre of the bubble at 120 pc, in the direction of the Sco-Cen supper shell as seen in HI at $(l, b) = (320°, 5°)$ (Heiles 1998). We note that this is not the centre of the Loop I at $(l, b) = (332°, 20°.7)$. Figure 3.34 shows a 3D projection of the field lines on the surface of the shell. From the figure, it is clear that the vantage point will define the appearance of the field lines for any observer.

Fig. 3.34 Direction of the
magnetic field lines on the
surface of a spherical shell
expanding in a magnetised
medium with an initial
uniform field parallel to the
Galactic plane. The field
lines will follow "meridian"
lines on the surface of the
shell as it is shown in the
figure. The Sun is located at
the origin of the three axis
and the units of are parsecs.
The radius of the shell is
120 pc and it is located at
120 pc from the Sun in the
direction $(l, b) = (320°, 5°)$

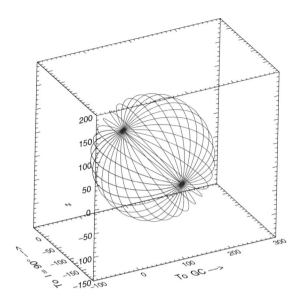

A Mollweide projection of the view of the field lines onto the sky is shown on top
in Fig. 3.35. The lines closer to the "poles" of the sphere have been masked to avoid
crowding. At the bottom, the same lines are drawn on top of the K-band polarisation
map. It is remarkable that this very simple model fits very well the direction of the
largest loops, above the Galactic plane.

Interestingly, if we plot the field lines for only half of the sphere, the hemisphere
of the closest "pole" of the shell, the direction of the field lines are more similar to
the magnetic field directions observed in *WMAP* K-band. In Fig. 3.36 we show this
comparison. On *top* are the predicted field line directions which are closest to the
Sun. At the *bottom* is shown the polarisation map with the field lines on top.

In this scenario, where the emission comes from the compression of the interstellar
field lines, one would expect emission coming from most of the surface of the shell.
This is clearly not the case. The NPS and the other large loops are the only places
where there is a significant amount of synchrotron. The reason for this difference in
synchrotron emission on different regions might be connected with the density of the
ISM at different sides of the shell. The interaction of the shell with a denser medium
might trap and accelerate cosmic rays more efficiently than in a thinner environment.
We can see how the gas density is distributed in the environment where the Sco-
Cen bubble lies. Lallement et al. (2003) presented a 3D map of the nearby ISM by
mapping the absorption of Na lines by neutral interstellar gas using 1000 lines-of-
sight. In Fig. 3.37 two sketches of the distribution of neutral gas in the vicinity of the
Sun are shown. Also, the location of the Sco-Cen bubble and some denser clouds are
marked. The section of the shell that has higher emission, corresponding to the NPS,
is close to the Ophiuchi clouds, one of the closest star forming complex at 139 pc

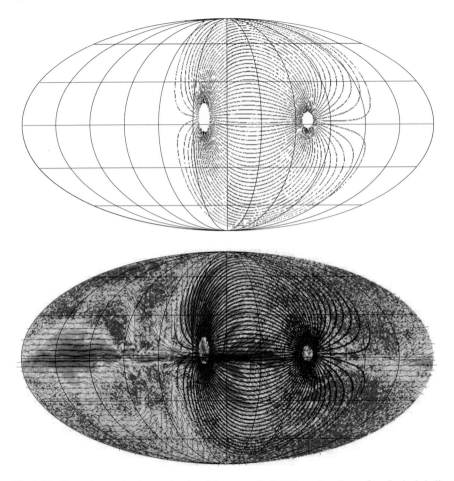

Fig. 3.35 On *top* is a projection on the sky of the magnetic field lines directions of a spherical shell of 120 pc of radius located at 120 pc in the direction. The *bottom panel* shows the same field lines drawn on *top* of *WMAP* K-band polarisation map. The *black* vectors on the *bottom* map show the polarisation angle at K-band. The grid spacing is 30° in both *l* and *b*

(Mamajek 2008). The larger density in this region might be the reason of the larger brightness of the NPS.

This model provides a relatively simple explanation for the polarisation angle distribution observed by *WMAP*. In Sect. 3.2, we showed that the polarisation angle along some filaments is systematically different from the direction defined by the polarisation intensity. This is also consistent with this view, in which the filaments are not limb-brightened structures, but the result of the illumination of some particular field lines due to more compression of the field or a larger density of CR.

Mertsch and Sarkar (2013) analysed the angular power spectrum of the diffuse synchrotron emission as observed in the Haslam et al. (1982) map. They compared

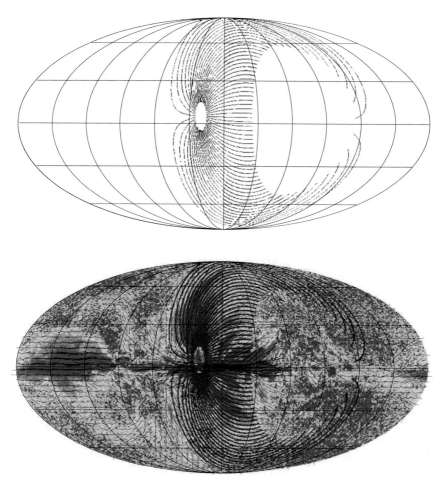

Fig. 3.36 Same as Fig. 3.35 but this time only one hemisphere of the field lines over the shell are drawn. We draw the lines of the hemisphere of the "pole" closest to the Sun. The grid spacing is 30° in both l and b

the observed power spectrum with a modelled one, which includes $O(1000)$ SN shells distributed in the Galaxy with scales ~ 100 pc. The inclusion of these SN shells greatly improves the fit of the power spectrum. If the results from their modelling are correct, these shell-like structures are quite common in the Galaxy, supporting the idea of an SN remnant origin for the diffuse synchrotron radiation.

We think that the model discussed here, which implies a nearby origin for Loop I and the largest filaments, describes the data better than the models connecting these large filaments with activity from the Galactic centre. One of the arguments that Bland-Hawthorn and Cohen (2003) uses to connect the NPS with nuclear activity is that the radio continuum emission from the NPS is thermal, originated by a shock induced UV field which ionized the gas. The spectral indices measured in this work

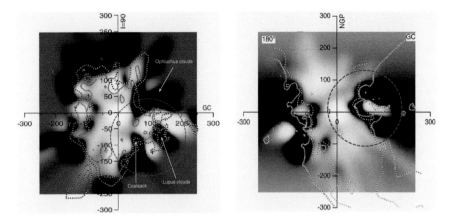

Fig. 3.37 Views of the nearby distribution of neutral gas from Lallement et al. (2003) where darker represents denser regions. On the *left* is a view from above the Galactic plane centred at the Sun location. On the *right* is a perpendicular view, this time with the vantage point on the plane of the Galaxy also centred at the Sun location. We have drawn the projection of the shell as a *red line*. The area of the shell where the NPS shines corresponds to the larger density region of the Ophiochi clouds marked on *left*

show the opposite, a steep spectral index consistent with diffuse synchrotron emission. Also, the polarisation observed is the signature of synchrotron and not free-free emission. In another work, Jones et al. (2012) claim that the magnetic field is perpendicular to the extension of the GC spur (Filament No. 1 in our analysis), therefore, being consistent with a toroidal magnetic field along that spur. As we have shown using the same *WMAP* data, this is not the case. The magnetic field lines are parallel to the extension of the filament, a fact which is unsuitable for a toroidal field along the spur. Carretti et al. (2013), with polarisation data at 2.3 GHz, detected two polarised radio lobes that encompass around 60° in the inner Galaxy. They claim that the origin of these lobes is connected with a large star formation activity in the central 200 pc of the Galaxy that can transports $\sim 10^{55}$ ergs of energy into the Galactic halo. This picture is valid for the emission from Filaments No. 1, 3, 8 and 9 from our analysis. We remark that this hypothesis is not incompatible with the existence of the local shell we have discussed here earlier.

There are still many questions regarding the polarised emission from the inner Galaxy. The connection with the *WMAP* haze and the *Fermi* bubbles observed by Su et al. (2010) is an open question. Also, the origin of the GC arc, which is located very close to the Galactic centre and the polarisation angle is remarkably well aligned along it. The absence of a corresponding filament below the plane raise doubts about a Galactic nucleus related origin.

Another very interesting feature visible in the *WMAP* polarisation amplitude maps at 1° resolution are thin filaments that rise symmetrically from the Galactic plane towards positive latitudes. We show their location Fig. 3.38. Their apparent width is ≈1°, which is equal to the resolution of the maps so probably their width is unresolved

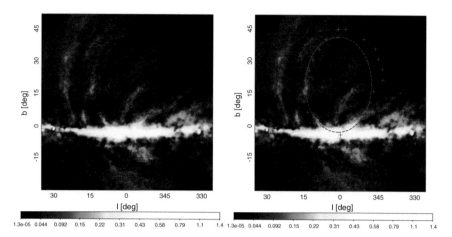

Fig. 3.38 Thin filaments visible at $1°$ scale on *WMAP* K-band polarisation map. *Left* and *right* show the same map with the aim of comparing the remarkably ellipsoidal shape of the thin shell. The *blue line* on the *right* shows an eyeball fit to the structure while the *magenta crosses* mark the borders of the *Fermi* north-bubble as defined in Su et al. (2010). We used an asinh colour scale to highlight the dim filament. The units of both maps are mK

in these maps. They closely follow the geometry of the northern *Fermi* bubble. There is no analogue below the Galactic plane. It is difficult to avoid the connection between these thin and very symmetric ellipsoidal structures, which extend up to $b \approx 45°$, which implies a linear size ~ 8 kpc if we assume a distance of 8 kpc to them. Little attention has been given to these filaments in the literature because they are relatively weak (here we highlight them using a non-linear scale in the figure). They lie in the depolarisation band of the low frequency ($\lesssim$ few GHz) polarisation maps so data from the *Planck* polarisation mission can shed light on the characteristics of these most probably limb-brightened features.

It is clear that the inner region of the Galaxy is a very interesting environment and to explain the origin of the polarised emission here one needs to take into account the physics of the local ISM as well as the phenomena occurring at the vicinities of the Galactic centre, as the associations we have shown suggest.

3.6.3 CMB Foreground Contribution

With the aim of quantifying the contribution of the polarised filaments in the context of foreground to the CMB, we have calculated the power spectra of the polarised synchrotron sky. Using masks which include/remove the observed filaments, we calculate the spectra and then compare these with the expected E and B-modes components of the CMB, to asses the level of contamination that the filaments induce at different angular scales.

To simplify the estimation, we used a simulated version of the polarised synchrotron sky using the *Planck* Sky Model (see Sect. 2.2). By doing this, we do not have to worry about the noise present in the observed *WMAP* polarisation maps. The simulated sky maps are created with an angular resolution of 1°. Because we are interested on the large angular scales, this value is small enough for our purpose. The simulated template is qualitatively similar to the polarised sky as seen by *WMAP*. In Fig. 3.39 we show the polarised intensity maps at K-band of the PSM and also *WMAP*, both with 1° resolution. There is a good agreement between both maps. The noise effect is visible on small angular scales in the *WMAP* map. The similarity between the simulation and the observed polarised synchrotron allows us to carry on with this estimation.

In order to test the contribution of the filaments to the power spectrum of the full sky, we calculated the EE and BB spectra for the simulated sky using two different masks. One that suppresses all the pixels on a 10° strip on the Galactic plane and a second mask that masks out the emission from the polarised features at high latitudes. These masks can be seen in Fig. 3.40.

The power spectrum was computed using the publicly available PolSpice package (Chon et al. 2004). This software measures the the angular auto- and cross-power spectra $C(l)$ of Stokes I, Q and U. It is well suited for our applications because it can correct for the effects of the masks, taking into account inhomogeneous weights

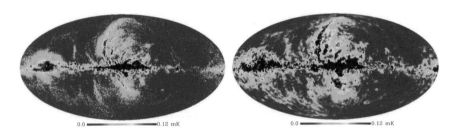

Fig. 3.39 Comparison between the *WMAP* K-band polarisation intensity map (*left*) and the PSM (*right*), both at 1° resolution. The maps are in antenna temperature units and are plotted using a linear colour scale

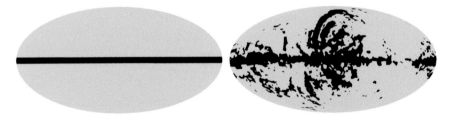

Fig. 3.40 The two mask used in the analysis. On the *left* is the one that masks the Galactic plane, all the pixels where $|b| \leq 5°$ and it covers $\approx 7\%$ of the sky. On the *right* is the one that we used to mask the emission from the filaments. This masks covers $\approx 27\%$ of the sky

given to the pixels of the map. The mixing of the E and B modes due to the masked sky and pixel weights is corrected for, so an unbiased estimate of the B-mode spectrum is given.

We also calculated a theoretical CMB power spectrum using the CAMB package (Lewis et al. 2000). For this, we used the cosmological parameters from the *WMAP* 9-year release, listed in Hinshaw et al. (2013). We used a tensor-to-scalar ratio $r = 0.01$ for the B-modes spectrum calculation because many inflationary models predict r to be close to this value. We calculated the power spectrum for two different frequencies. First, at 20 GHz, where the polarised synchrotron is expected to dominate and also at 80 GHz, where its contribution is more reduced. Because we aim to estimate the contribution of the synchrotron filaments, we do not include any dust contribution at these higher frequencies. We used a spectral index of $\beta = -3.1$ to scale the emission from 20 to 80 GHz. This value is an average for the high latitude filaments as we have measured previously in this chapter. We note that β_{20-80} might be steeper than the -3.1 value that we used due to frequency steepening of the spectrum. If this occurs, the contribution that we show here at 80 GHz will be an overestimate for the power of the synchrotron filaments at that 80 GHz.

In Fig. 3.41 we show the resulting polarisation power spectra for the two different frequencies. The black lines in both plots show the expected E and B-modes of the CMB power spectra. The red lines show the E and B-modes of the polarised synchrotron template computed using only the Galactic plane mask, while the blue lines show the spectra computed using the filaments mask. The difference between the spectra calculated using these two masks is more pronounced at the large angular scales (low multipole value), as expected given the extension of the filaments. The B-mode power calculated using the filament mask is ~ 140 times smaller for

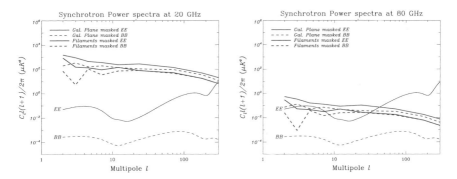

Fig. 3.41 Power spectra for the polarised synchrotron component at two frequencies compared to the theoretical CMB E and B-modes spectra. In *black* are the CMB spectra where the B-modes were calculated using a tensor-to-scalar ratio value of $r = 0.01$. In *colours* are the spectra of the synchrotron polarisation map using the two masks showed in Fig. 3.40. In *red* is the B-mode and E-mode spectrum calculated using the Galactic plane mask and in blue are the spectra computed using the filament mask. On the *left* is the expected contribution at 20 GHz and on the *right* are the spectra for the synchrotron component at 80 GHz

$\ell = 3$ than the power estimated using only the Galactic plane mask. With increasing multipole value, the difference of the spectra calculated with the different masks is smaller, reaching an steady value around $\ell = 60$ of 2 times, i.e. the foreground contribution of the filaments on B-modes is 2 times larger than when they are masked out.

The large difference at low-ℓ on the B-modes spectrum is important because the "low-ℓ bump" of the CMB B-modes spectrum will be a target for the experiments that aim to detect the cosmological B-mode signal (see e.g. Katayama and Komatsu 2011). This imply that this experiment will have to use large masks like the one used here or characterise precisely the synchrotron emission to be able to perform a clean removal of the emission from the maps. It is evident that foreground cleaning would be essential to characterise a cosmological B-mode signal in the low-ℓ range.

References

Alves, M. I. R., et al. (2012). A derivation of the free-free emission on the Galactic plane between l= 20°. and 44°. *MNRAS, 422*, 2429–2443.

Bennett, C. L., et al. (2013). Nine-year Wilkinson Microwave Anisotropy Probe (WMAP) observations: Final maps and results. *ApJS, 208(20)*, 20.

Berkhuijsen, E. M. (1973). Galactic continuum loops and the diameter-surface brightness relation for supernova remnants. *A & A, 24*, 143–147.

Berkhuijsen, E. M., Haslam, C. G. T., & Salter, C. J. (1971). Are the galactic loops supernova remnants? *A & A, 14*, 252–262.

Bingham, R. G. (1967). "Magnetic fields in the galactic spurs". In: MNRAS 137, p. 157.

Bland-Hawthorn, J., & Cohen, M. (2003). The large-scale bipolar wind in the Galactic center. *ApJ, 582*, 246–256.

Borka, V. (2007). Spectral indices of Galactic radio loops between 1420, 820 and 408 MHz. *MNRAS, 376*, 634–644.

Breitschwerdt, D., McKenzie, J. F., & Voelk, H. J. (1991). Galactic winds. I: Cosmic ray and wave-driven winds from the Galaxy. *A & A, 245*, 79–98.

Brown, R., Davies, R. D., & Hazard, C. (1960). A curious feature of the radio sky. *The Observatory, 80*, 191–198.

Bunner, A. N. et al. (1972). Soft X-Rays from the Vicinity of the North Polar Spur. In: ApJ 172, p. L67.

Burn, B. J. (1966). On the depolarization of discrete radio sources by Faraday dispersion. In: MNRAS 133, p. 67.

Carretti, E., et al. (2013). Giant magnetized outflows from the centre of the Milky Way. *Nature, 493*, 66–69.

Chon, G., et al. (2004). Fast estimation of polarization power spectra using correlation functions. *MNRAS, 350*, 914–926.

Colomb, F. R., Poppel, W. G. L., & Heiles, C. (1980). Galactic HI at –b– ¿= 10°. II. Photographic presentation of the combined Southern and Northern data. *A & AS, 40*, 47–55.

Davies, R. D., et al. (2006). A determination of the spectra of Galactic components observed by the Wilkinson Microwave Anisotropy Probe. *MNRAS, 370*, 1125–1139.

Egger, R. J. (1995). "Loop I—an Active Superbubble". In: The Physics of the Interstellar Medium and Intergalactic Medium. Ed. by A. Ferrara et al. Vol. 80. Astronomical Society of the Pacific Conference Series, p. 45.

Egger, R. J., & Aschenbach, B. (1995). Interaction of the Loop I supershell with the Local Hot Bubble. *A & A, 294*, L25–L28.

Finkbeiner, D. P. (2004). Microwave interstellar medium emission observed by the Wilkinson Microwave Anisotropy Probe. *ApJ, 614*, 186–193.

Frail, D. A. et al. (1996). "The Pulsar Wind Nebula around PSR B1853+01 in the Supernova Remnant W44". In: ApJ 464, p. L165.

Gold, B., et al. (2011). Seven-year Wilkinson Microwave Anisotropy Probe (WMAP) observations: Galactic foreground emission. *ApJS, 192*(15), 15.

Guzmán, A. E., et al. (2011). All-sky Galactic radiation at 45 MHz and spectral index between 45 and 408 MHz. *A & A, 525*(A138), A138.

Haslam, C. G. T. et al. (1982). "A 408 MHz all-sky continuum survey. II—The atlas of contour maps". In: A & AS 47, p. 1.

Heiles, C. (1974). "An almost complete survey of 21 CM line radiation for —b— i= 10°. II. The accurate data on machine readable magnetic tape". In: A & AS 14, p. 557.

Heiles, C. (1984). HI shells, supershells, shell-like objects, and 'worms'. *ApJS, 55*, 585–595.

Heiles, C. (1998). "TheMagnetic Field Near the Local Bubble". In: IAU Colloq. 166: The Local Bubble and Beyond. Ed. by D. Breitschwerdt, M. J. Freyberg, & J. Truemper. Vol. 506. Lecture Notes in Physics. Springer, Berlin, pp. 229–238.

Heiles, C., et al. (1980). A new look at the north Polar Spur. *ApJ, 242*, 533–540.

Hinshaw, G., et al. (2013). Nine-year Wilkinson Microwave Anisotropy Probe (WMAP) observations: Cosmological parameter results. *ApJS, 208*(19), 19.

Jones, D. I., et al. (2012). Magnetic substructure in the Northern Fermi Bubble revealed by polarized microwave emission. *ApJ, 747*(L12), L12.

Jun, B.-I., & Jones, T. W. (1999). Radio emission from a Young Supernova Remnant interacting with an interstellar cloud: Magnetohydrodynamic simulation with relativistic electrons. *ApJ, 511*, 774–791.

Kalberla, P. M. W., et al. (2005). The Leiden/Argentine/Bonn (LAB) survey of Galactic HI. Final data release of the combined LDS and IAR surveys with improved stray-radiation corrections. *A & A, 440*, 775–782.

Katayama, N., & Komatsu, E. (2011). Simple foreground cleaning algorithm for detecting primordial B-mode polarization of the cosmic microwave background. *ApJ, 737*(78), 78.

Kogut, A., et al. (2007). Three-Year Wilkinson Microwave Anisotropy Probe (WMAP) observations: Foreground polarization. *ApJ, 665*, 335–362.

Koo, B.-C., Heiles, C., & Reach, W. T. (1992). Galactic worms. I—Catalog of worm candidates. *ApJ, 390*, 108–132.

Lallement, R., et al. (2003). 3D mapping of the dense interstellar gas around the Local Bubble. *A & A, 411*, 447–464.

Landecker, T. L., et al. (1999). DA 530: A Supernova Remnant in a Stellar Wind Bubble. *ApJ, 527*, 866–878.

Large, M. I., Quigley, M. F. S., & Haslam, C. G. T. (1966). "A radio study of the north polar spur. II. A survey at low declinations". In: MNRAS 131, p. 335.

Large, M. I., Quigley, M. J. S., & Haslam, C. G. T. (1962). "A new feature of the radio sky". In: MNRAS 124, p. 405.

Lewis, A., Challinor, A., & Lasenby, A. (2000). Efficient computation of cosmic microwave background anisotropies in closed Friedmann-Robertson-Walker models. *ApJ, 538*, 473–476.

López-Caniego, M., et al. (2009). Polarization of the WMAP point sources. *ApJ, 705*, 868–876.

Mamajek, E. E. (2008). "On the distance to the Ophiuchus star-forming region". In: Astronomische Nachrichten 329, p. 10.

Mathewson, D. S., & Ford, V. L. (1970). Polarization measurements of stars in the Magellanic Clouds. *AJ, 75*, 778–784.

Mertsch, P., & Sarkar, S. (2013). Loops and spurs: The angular power spectrum of the Galactic synchrotron background. *Journal of Cosmology and Astroparticle Physics, 6*(041), 41.

Milogradov-Turin, J., & Uroiševič, D. (1997). Geometry of large radio loops at 1420 MHz. *Bulletin Astronomique de Belgrade*, *155*, 41–45.

Oppermann, N., et al. (2012). An improved map of the Galactic Faraday sky. *A & A*, *542*(A93), A93.

Page, L., et al. (2007). Three-Year Wilkinson Microwave Anisotropy Probe (WMAP) observations: Polarization analysis. *ApJS*, *170*, 335–376.

Planck Collaboration et al. (2013). "Planck intermediate results. IX. Detection of the Galactic haze with Planck". In: A & A 554, A139, A139.

Preibisch, T., & Mamajek, E. (2008). "The Nearest OB Association: Scorpius-Centaurus (Sco OB2)". In: Handbook of Star Forming Regions, Volume II. Ed. by B. Reipurth, p. 235.

Puspitarini, L., & Lallement, R. (2012). Distance to northern high-latitude HI shells. *A & A*, *545*(A21), A21.

Quigley, M. J. S., & Haslam, C. G. T. (1965). Structure of the radio continuum background at high Galactic latitudes. *Nature*, *208*, 741–743.

Reynolds, S. P., Gaensler, B. M., & Bocchino, F. (2012). "Magnetic Fields in Supernova Remnants and Pulsar-Wind Nebulae". In: Space Sci. Rev. 166, pp. 231–261.

Reynoso, E. M. et al. (1997). "A VLA Study of the Expansion of Tycho's Supernova Remnant". In: ApJ 491, p. 816.

Sofue, Y. (1977). Propagation of magnetohydrodynamic waves from the galactic center: Origin of the 3-kpc arm and the North Polar Spur. *A & A*, *60*, 327–336.

Sofue, Y. (1988). Vertical radio structures out of the Galactic plane and activities of the Galaxy. *PASJ*, *40*, 567–579.

Sofue, Y., Hamajima, K., & Fujimoto, M. (1974). "Radio Spurs and Spiral Structure of the Galaxy, II. On the Supernova Remnant Hypothesis for Spurs". In: PASJ 26, p. 399.

Sofue, Y., Reich, W., & Reich, P. (1989). The Galactic center spur: A jet from the nucleus? *ApJ*, *341*, L47–L49.

Spoelstra, T. A. T. (1971). "A Survey of Linear Polarization at 1415 MHz. II. Discussion of Results for the North Polar Spur". In: A & A 13, p. 237.

Spoelstra, T. A. T. (1972). "A Survey of Linear Polarization at 1415 MHz. IV. Discussion of the Results for the Galactic Spurs". In: A & A 21, p. 61.

Spoelstra, T. A. T. (1973). "Galactic Loops as Supernova Remnants in the Local Galactic Magnetic Field". In: A & A 24, p. 149.

Strong, A. W., Orlando, E., & Jaffe, T. R. (2011). The interstellar cosmic-ray electron spectrum from synchrotron radiation and direct measurements. *A & A*, *534*(A54), A54.

Su, M., Slatyer, T. R., & Finkbeiner, D. P. (2010). Giant gamma-ray bubbles from Fermi-LAT: Active Galactic nucleus activity or bipolar Galactic wind? *ApJ*, *724*, 1044–1082.

Tomisaka, K. (1992). The evolution of a magnetized superbubble. *PASJ*, *44*, 17–191.

Tsuboi, M., et al. (1995). Galactic center arc–polarized plumes complex at 43 GHz. *PASJ*, *47*, 829–836.

van der Laan, H. (1962). "Expanding supernova remnants and galactic radio sources". In: MNRAS 124, p. 125.

Weaver, H. (1979). "Large supernova remnants as common features of the disk". In: The Large-Scale Characteristics of the Galaxy. Ed. by W. B. Burton. Vol. 84. IAU Symposium, pp. 295–298.

Weaver, R., et al. (1977). Interstellar bubbles. II. Structure and evolution. *ApJ*, *218*, 377–395.

Wolleben, M. (2007). A new model for the loop I (North Polar Spur) region. *ApJ*, *664*, 349–356.

Wolleben, M., et al. (2006). An absolutely calibrated survey of polarized emission from the Northern sky at 1.4 GHz. Observations and data reduction. *A & A*, *448*, 411–424.

Chapter 4
QUIET Galactic Observations

The Q/U Imaging ExperimenT (QUIET) was a project that aimed to measure the E and B-mode polarisation by simultaneously measuring the Stokes parameters Q and U, hence its name. Located at 5080 m altitude in the Atacama desert in Chile, it observed six regions of the sky at 43 and 95 GHz between October 2009 until December 2010. Four of these regions are the "CMB patches", which are located in regions with low Galactic emission. The other two regions corresponds to the "Galactic patches", which lie on the Galactic plane.

Here we describe the instrument, observations and data processing. Then, we describe a test that we performed on a CMB field for residual foreground contamination and systematic effects. The maps from the Galactic observations are described. These maps are going to be the base of a paper from the collaboration and here we show some ongoing analysis that will be part of that work.

4.1 The Instrument

The instrument is fully described in QUIET Collaboration et al. (2012b). Here we provide a basic summary of its main characteristics.

QUIET observed the polarised millimetre-wavelength sky at 43 and 95 GHz using arrays of correlation polarimeters mounted on a side-fed telescope. The ℓ range to which QUIET is sensitive includes the first three peaks of the CMB E-mode spectrum. The instrument was located at the Chajnantor plateau in the Atacama desert in Chile. Extreme dryness and altitude, over 5080 m, makes this location ideal for observations in the GHz frequency range. QUIET uses the facilities and telescope mount of the former Cosmic Background Imager (CBI) (Taylor et al. 2011; Padin et al. 2002), a 13 element interferometer that observed until June of 2008. The CBI mount allows rotation in azimuth, elevation and around the optical axis (deck angle). QUIET is enclosed in an absorbing screen to minimise ground spillover. Figure 4.1 shows a diagram of the instrument alongside a photograph at the observing site.

© Springer International Publishing Switzerland 2016
M. Vidal Navarro, *Diffuse Radio Foregrounds*, Springer Theses,
DOI 10.1007/978-3-319-26263-5_4

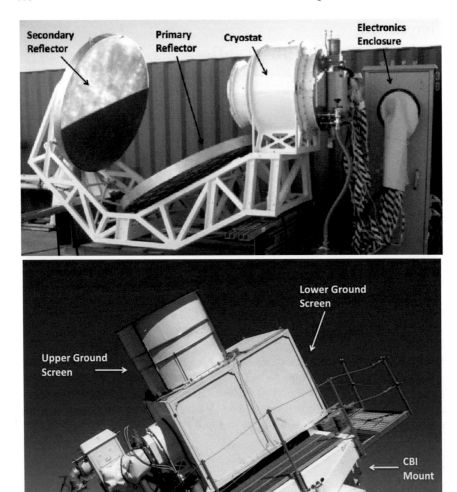

Fig. 4.1 *Top* The QUIET telescope, before being located in the absorbing enclosure and mount, with the arrows showing the main components. *Bottom* The instrument after being mounted on the former CBI mount. These images are from QUIET Collaboration et al. (2012b)

QUIET uses a 1.4 m diameter Dragonian antenna (Dragone 1978), in which a paraboloidal main reflector illuminates a concave hyperboloidal subreflector. This design has a number of characteristics that are ideal for CMB-polarisation instruments. It has excellent polarisation characteristics that minimise the instrument polarisation and beam distortions, minimal sidelobes and a wide field-of-view, which is

7° diameter in this case. A detailed description of the optical components of QUIET is given in Imbriale et al. (2011).

QUIET operates in two frequency bands: Q–band, centred at 43 GHz and W–band, centred at 95 GHz using two arrays of detectors. The Q–band array contains 19 modules, arranged in a hexagonal pattern. 17 measure polarisation and the remaining 2 measure the CMB temperature anisotropy. The W–band array, operating at 95 GHz consists of 90 modules, 84 measuring polarisation and 6 measuring the CMB temperature. All the modules are based on High Electron Mobility Transistors (HEMTs) that provide simultaneous measurements of Stokes Q and U, in a very compact configuration, with a foot-print of 5.08 cm $\times$ 5.08 cm at Q–band and 3.18 cm $\times$ 2.90 cm at W–band. Each array has a dedicated cryostat where the modules operate at 20 K. These two arrays are the largest HEMT-based arrays ever built. Figure 4.2 shows a photograph of the W–band array. The 43 GHz array, has the best sensitivity ($69\,\mu K\sqrt{s}$) and the lowest instrumental systematic errors ever achieved in this band, which contribute to the tensor-to-scalar ratio at $r < 0.1$. The W–band array, also present the lowest systematic error to date and it contributes to $r < 0.01$.

Each element of a given array is a correlation polarimeter which can measure simultaneously both Stokes Q and U. Before each module in the optical chain, a septum polariser converts the orthogonal incident components E_x and E_y into

Fig. 4.2 The QUIET W–band polarimeter array. This is the largest HEMT-based array ever assembled to-date

left-circularly polarised component $L = (E_x + iE_y)/\sqrt{2}$ and a right-circularly polarised component $R = (E_x - iE_y)/\sqrt{2}$. The septum spatial orientation is used to define the instrumental position angle. The L and R outputs from the septum are feed into the module.

In the pseudo-correlator receivers, the output is a product of gain terms. The modules receives L and R as input and the Stokes parameters are measured in the following way:

$$I = |L|^2 + |R|^2,$$
$$Q = 2\,\mathrm{Re}(L^*R),$$
$$U = -2\,\mathrm{Im}(L^*R),$$
$$V = |L|^2 - |R|^2, \tag{4.1}$$

where $*$ denotes the complex conjugation. V is expected to be zero and is not measured by the instrument. We can see that by using circular feeds, the output Q, U Stokes parameters are constructed using the product of the two circular feeds. This is an advantage over using linear feeds, where the Q Stokes parameter is constructed by subtracting two auto-correlations from each other, i.e. $Q = \langle E_x^2 \rangle - \langle E_y^2 \rangle$. This operation is inherently more noisy than the product, as in the case of circular feeds, because involves the subtraction of two large quantities to produce a small number. In general, circular feeds are better for linear polarisation measurements and linear feeds are better for circular polarisation measurements.

Figure 4.3 shows a schematic and a picture of the QUIET module, where L and R, coming from the septum polariser, traverse separate amplification "legs". A phase switch in each leg modulates the signal at 4 kHz or 50 Hz. After amplification and modulation, the signals are combined in a 180° hybrid coupler which, for voltage inputs a and b, yields $(a + b)/\sqrt{2}$ and $(a - b)/\sqrt{2}$ as outputs. The outputs of the hybrid coupler are split, half of each goes to detector diodes D_1 and D_4 respectively. The other halves of the output powers go to a 90° coupler which, for voltage inputs a and b, yields $(a + ib)/\sqrt{2}$ and $(a - ib)/\sqrt{2}$ as outputs. The outputs of this coupler are detected in diodes D_2 and D_3.

After the diodes have detected the sky signal, the Readout system digitises the output of each four of them at a rate of 800 kHz. Continuous blocks of 125 μs in length are subtracted to produce the polarisation sensitive streams. Averaging over 10 ms the blocks produces a stream proportional to Stokes I for all the diodes. The average and demodulation therefore return I and Q or U respectively for each diode. Table 4.1 shows the idealised detector diode outputs.

The choice of circularly-polarised inputs thus allows the measurement of Q and U Stokes parameters simultaneously, giving an advantage in detector sensitivity. This characteristic is another advantage over bolometer receivers (such as those used in the high frequency instrument in *Planck*), where each one only measures a single Stokes parameter.

In Table 4.2 are listed the main characteristics of the QUIET instrument which summarise this description.

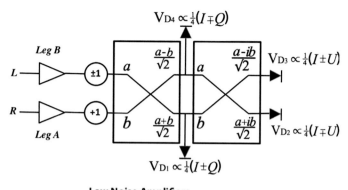

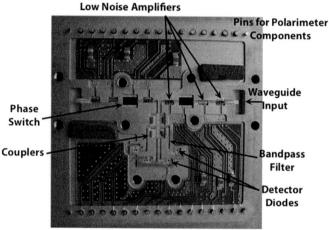

Fig. 4.3 *Top* Schematic of the signal flow in the QUIET module. Two circularly polarised inputs L and R are amplified by HEMP-based low noise amplifiers. Phase switches "± 1" provide electronic modulation. The signal are combined in hybrid couplers and detected at diodes D_1 to D_4. *Bottom* Picture of a 5 cm $\times$ 5 cm Q–band module

Table 4.1 Idealized detector diode outputs for a polarimeter assembly

Diode	Raw output	Average	Demodulated
D_1	$\propto \frac{1}{4}(I \pm Q)$	$\propto \frac{1}{4}I$	$\propto \frac{1}{2}Q$
D_2	$\propto \frac{1}{4}(I \mp U)$	$\propto \frac{1}{4}I$	$\propto -\frac{1}{2}U$
D_3	$\propto \frac{1}{4}(I \pm U)$	$\propto \frac{1}{4}I$	$\propto \frac{1}{2}U$
D_4	$\propto \frac{1}{4}(I \mp Q)$	$\propto \frac{1}{4}I$	$\propto -\frac{1}{2}Q$

4.2 Observations

In this Section the QUIET observations are described. I spent three weeks at the observing site during 2010, executing the daily tasks required to keep the telescope running. These tasks included preparing the observation scripts; keeping track of the

Table 4.2 QUIET instrument overview

Band	Q	W
Frequency (GHz)	43	95
Bandwidth (GHz)	7.6	10.7
No. of polarisation assemblies	17	84
No. of temperature anisotropy assemblies	2	6
FWHM angular resolution (arcmin)	27.3	11.7
Field-of-View	7°	8°.1
ℓ range	≈25−475	≈25−950
Instrument sensitivity ($\mu K \sqrt{s}$)	69	87
Averaged module sensitivity ($\mu K \sqrt{s}$)	275	756

observations, recording on a daily basis the data onto blue-ray DVD disks (standard hard drives are prone to fail at the high altitude site and optical disks proved to be a reliable and economic way of downloading the data from the observatory); performing a nightly optical pointing procedure once a week; and helping to fix any hardware problem, as well as the routinely helium refill for the cryostat.

QUIET observed in four 15° × 15° CMB fields. Their location and size were chosen in order to satisfy certain conditions. They have to be separated in RA in order to observe them in sequence during the 24 four hours of a day. They also have to transit at an elevation larger than 43°, which is the mechanical lower limit of the telescope. They should present minimal foreground contamination, which is assessed using *WMAP* data. Finally, they have to be at least 30° away from the Sun and the Moon during the observations. Two additional fields in the Galactic plane were observed to constrain the properties of the foregrounds with high signal-to-noise ratio. Figure 4.4 shows the position in the sky of the fields. Table 4.3 lists the equatorial coordinates of the centre of each patch and the observing time with both Q- and W-band arrays.

The observation of the patches were performed by fixing the telescope elevation and scanning in azimuth, with an amplitude of 15°, while the patch moves along the field-of-view of the instrument due to Earth rotation. To start, the telescope is positioned ∼3°.5 (one focal plane radius) ahead of the edge of the patch. Then, it scans with a mean azimuth speed of ∼5° s^{-1}. The duration of these *constant elevation scans* (CES) is between 40 and 90 min. After this time, when the field has drifted about 15° in the sky, the telescope is re-pointed to start a new CES. This scanning strategy provides a simple way to subtract the atmospheric contribution, proportional to the atmospheric opacity, which is fairly constant at a fixed elevation. By observing the same field at different times of the day, a same region in the sky is observed with

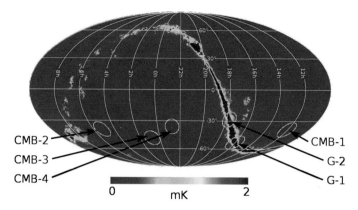

Fig. 4.4 QUIET fields in Equatorial coordinates, superimposed on the *WMAP* Q–band map

Table 4.3 Patch locations and integration times

Patch	RA (J2000)	Dec. (J2000)	l (deg)	b (deg)	Integration Q (h)	Integration W (h)
CMB-1	12^h04^m	$-39°00'$	292.9	22.9	905	1830
CMB-2	05^h12^m	$-39°00'$	243.2	-35.3	703	1416
CMB-3	00^h48^m	$-48°00'$	304.5	-69.1	837	1375
CMB-4	22^h44^m	$-36°00'$	6.8	-61.6	223	631
G-1	17^h46^m	$-28°56'$	0.0	0.0	92	324
G-2	16^h00^m	$-53°00'$	330.0	0.0	311	697

Note that the RA of the fields distributed over the 24 h, allowing continuous observations. The CMB fields are located away from the Galactic plane, in order to minimise foreground contamination

different parallactic angle.[1] The parallactic coverage is increased by weekly deck rotations (the entire instrument rotates about the boresight).

In addition to the science targets, a number of additional observations are performed on a daily and weekly basis to calibrate the data. These are discussed in the following section. In Table 4.4 we list the distribution of observation time among the different types of observations.

4.2.1 Calibration

In order to obtain calibrated maps of the sky and power spectra, from the instrument measurements, four quantities are required. They are the detector responsivities to translate from detector voltages to antenna temperature, the detector polarisation

---

[1] The spherical angle between the hour circle of the observed object and the great circle that connects the object and the zenith.

Table 4.4 Summary of QUIET observations

Band	Q	W
Season start	24/10/2008	12/08/2009
Season end	13/06/2009	22/12/2010
Total observing hours	3458	7493
CMB observing (%)	77	72
Galactic observing (%)	12	14
Calibration (%)	7	13
Other (%)	4	1

Table 4.5 Regular calibration observations

Source	Duration (min)	Frequency
Sky dips	3	every 1.5 h
Tau A	20	every 1–2 days
Moon	60	weekly
Jupiter	20	weekly
Venus	20	weekly
RCW38	20	weekly

angles, the beam profiles and a pointing model. They are obtained by the observation of sources with well-known characteristics. These calibrators are listed in Table 4.5, along with the frequency in which they are observed.

In order to obtain the responsivity of the modules, the Crab nebula (Tau A) with a polarised flux density 22.1 ± 0.6 Jy at 41 GHz (uncertainty $\approx 2.6\%$) and a spectral index $\alpha = -0.350 \pm 0.026$ as measured by *WMAP* (Weiland et al. 2011), is observed every two nights. The observations are performed at four parallactic angles (varying the deck angle). This produces a sinusoidal modulation of the polarised signal which can be fitted to obtain the responsivity for each detector diode within the module as shown in Fig. 4.5. This source is observed with the central horn of the array and it provides the fiducial absolute responsivity for the rest of the array. The responsivities of the other modules are measured using the *sky dips*. These consist of observations at fixed azimuth with a $\pm 3°$ amplitude in elevation. They generate a large (~ 100 mK) signal which allow a calibration of the total power response. The ratio between the TP and polarised responsivity is stable within a few per cent and these sky dips are performed before each CES to quantify the responsivity response of array.

The polarisation angle of the detectors is determined using Moon observations. They produce a quadrupolar polarised pattern, the orientation of which determines the orientation for each detector. Figure 4.5e shows a polarisation map of the Moon from one module. The dotted line in the figure defines the polarisation orientation of the detector.

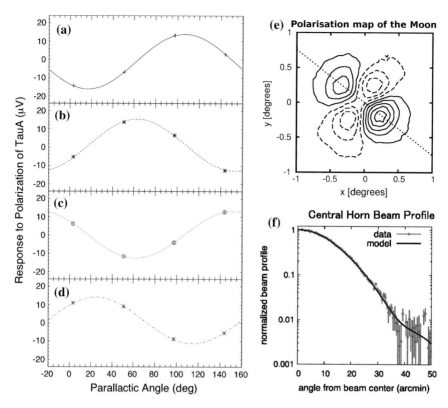

Fig. 4.5 **a, b, c, d** QUIET central horn polarimeter responsivity for Tau A at four deck angles. The horizontal axis represents the position angle of the focal plane in equatorial coordinates. Each panel shows the measurement from a different detector diode, which corresponds to Q, $-U$, U and $-Q$ respectively (see Table 4.1). The uncertainty in each measurement is smaller than the size of the points. **e** Polarisation map of the Moon from one detector diode, where the dotted line shows the polarisation angle of the detector. **f** The polarisation beam profile of the central horn, constructed using observations of Tau A. These figures are taken from QUIET Collaboration et al. (2011)

The beam profiles are obtained from maps created using the observations of Tau A and Jupiter. The Q–band horns produce a beam with an average FWHM of 27′.3. The elongation is small, of order 1 % and it is reduced thanks to the weekly deck rotations and diurnal sky rotation. One-dimensional symmetrised beam profiles are calculated and modelled in Monsalve (2010). The polarisation beam profile from Tau A observations is shown Fig. 4.5f. The model is a good fit to the data and the beam is very well characterised within the FWHM of the beam.

The pointing is obtained using observations of the Moon, Jupiter and some stars. Moon observations are used to measure the position of the individual beams in the sky for each horn of the array. These data are used to generate a model that relates the beam position in the sky with respect to telescope coordinates. Weekly, a set of stars are observed using an optical telescope mounted in the QUIET deck. These data are used to produce a mechanical model of the mount. Additionally, Jupiter observations from QUIET are included in the generation of the model. The rms pointing for the observations at Q–band is $3'.5$.

4.3 Data Reduction Pipeline and Map Making

QUIET uses two independent pipelines to process and analyse the data. Pipeline A is based on a pseudo-C_ℓ method which has been used by a number of other experiments (e.g. Brown et al. 2009; Chiang et al. 2010). It is computationally efficient, so it allows the execution of a set of blind tests to characterise systematic errors. Pipeline B is based on a maximum-likelihood framework (see e.g. Tegmark 1997; Bond et al. 1998) and it has the advantage of producing unbiased maps with the full noise-covariance matrix. This pipeline is computationally expensive so a reduced number of tests are done to the data. Both pipelines select the good data, characterise the noise in the frequency domain and filter the data in order to remove baseline level, high-frequency spikes and ground contamination.

Here we focus on the process of Pipeline B, because it will produce the maps of the Galactic patches that we use in the rest of the analysis. We focus in particular on the filtering of the Galactic data, as it is different from the filtering of the data from the CMB patches. This is because the Galactic plane demonstrates strong emission that generates artefacts if the standard (for CMB data) map-making is applied.

4.3.1 Map Making

To obtain maps of the sky from the QUIET data, the first step is the *pre-processing*, which corrects for a non-linearity detected in the hardware. Also in this step, the receiver data is synchronised with the telescope pointing information. Then, there is a *noise modelling* step in which the power spectral density of the noise (Eq. 2.5) is fitted using a three-parameter model: an amplitude for the white noise, the $1/f$ knee frequency and the slope of the $1/f$ component of the noise. A last step is the *filtering* in which systematic features in the noise power spectra that are not accounted for by the model are subtracted from the data.

Filtering is important as it helps to remove possible contamination in the data. This contamination can be of diverse origin: atmospheric fluctuations, ground spillover or instrumental noise. The suppression of signal at different temporal frequencies is reflected in a decrease of power for some angular modes in the sky maps. The filters applied to the data are described now.

Apodized bandpass filter. This can be separated in two filters: a high-pass and a low-pass filter. The first one removes modes with frequencies lower than $f_{min} = x$ times the scan frequency (45 to 100 mHz). This is needed to suppress scan-synchronous contamination. The low-pass filter removes frequencies higher

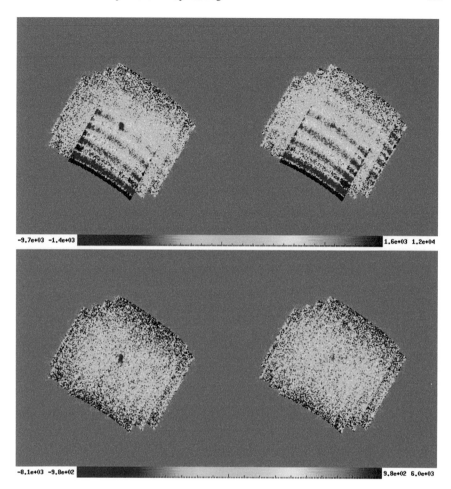

Fig. 4.6 High-pass filter effect. The panel on *top* shows Q and U unfiltered maps from a CES in the G-1 patch, using (mK) units. There are strips with high emission along the scanning direction. Their origin is ground contamination. The *lower panel* shows the same maps after filtering the modes with frequency higher than 0.05 times the scanning frequency

than 4.5 Hz. This is to eliminate narrow spikes that were discovered in the data. Figures 4.6 and 4.7 show the effect of these filters in maps of the Galactic centre patch G-1. These filters have to be applied to the Galactic data, so some spatial modes are lost in the final maps.

Azimuth filter. This is designed to remove the ground emission, suppressing any azimuthal structure remaining after adding CESs. If a strong source is present in the field, it adds artefacts along the scanning direction. This is shown in Fig. 4.8 as stripes around the Galactic centre. By fine-tuning the parameters of this filter and by applying it only to the worst CESs (the ones that show more ground contamination)

Fig. 4.7 Low-pass filter effect. On the *top panel* are the unfiltered Q and U maps of a constant elevation scan (CES) of patch G-1, using (mK) units. The filtered maps are in the *lower panel*. This filter removes high frequency modes along the scanning direction, resulting in a "smoothing" of the maps in that direction. This can be appreciated comparing the central negative source on the Q maps on the *left*. The unfiltered map on *top* shows a better angular resolution in this case

we managed to correct for the negative effects of this filter in the Galactic maps. Figure 4.9 shows two CES (*top* and *bottom*), one with the filtered producing artefacts and the second with the correct parameters that produce a cleaner map. For the Galactic patches, we selected individually all the CESs that require this filter in order to produce the final maps.

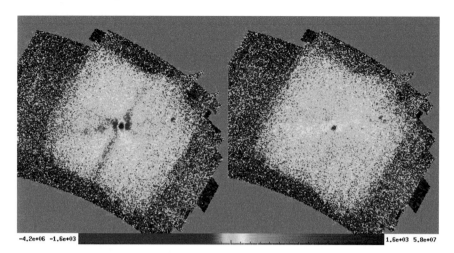

Fig. 4.8 Azimuth filter effect. Q and U filtered maps from the G-1 patch. The artefacts we see in the map on the Q map on the *left* are due to the azimuthal filtering of data, when a strong sources (such as the Galactic centre) is present. The U map on the *right* does not show such a problem as the central source is not very bright in Stokes U at Q–band

4.4 Test on Gaussianity on a CMB Field

In order to check the CMB maps for any residual foreground contamination or unwanted systematics, a simple test can be done. If the Stokes Q and U maps contain only CMB emission plus noise, the distribution of polarisation angles in the map should be Gaussian on average. We test this hypothesis, by building a histogram with the angles distribution in a CMB map from QUIET 95 GHz observations. The CMB-1 field shows some foreground synchrotron contamination at 43 GHz (see Sect. 4.5), so we tested for any residual contaminant in the 95 GHz data. Figure 4.10 shows the Stokes Q and U maps from the CMB-1 field. These maps are clearly dominated by large scale modes that have to be filtered before measuring the distribution of polarisation angles.

We first tried filtering these maps using an apodizing filter in multipole space. To do this, we first calculated the polarisation power spectrum of the data using the IANAFAST routine from the HEALPIX package. Then, we multiplied the spectrum by a filter with a shape similar to the arctangent function in order to remove the large scale modes from the spectrum. In Fig. 4.11 we show the EE spectrum and the filter used. We then reconstructed the polarisation maps from the filtered $a_{\ell,m}$. These maps are shown in Fig. 4.12. Most of the large scale emission present in the maps from Fig. 4.10 is removed. However, some additional artefacts are visible near the edges and also as a "dipolar" pattern at the centre of the maps. These anomalies are caused by edge-effects during the spherical harmonics decomposition of the maps. They arise as these maps are only a fraction of the celestial sphere, as opposed to full-sky maps, where the decomposition in multipoles is well defined.

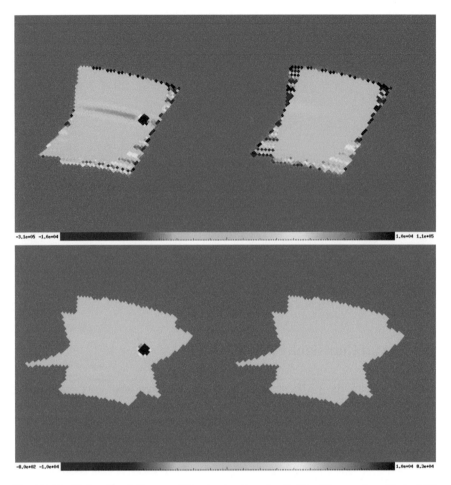

Fig. 4.9 *Top* Stokes Q and U maps of Tau A using the azimuth filter. The negative effects of this filter are corrected by the fine-tuning the filter parameters and applying it only to the CESs that show ground contamination. The result of the correct application of the filter is shown in the *bottom panel*. The strips along the scanning direction that cross the source disappear

A better way to account for the contaminated modes is to filter the maps during the map-making procedure. This is called eigenmode filtering and it takes into account the proper noise weighting, giving a large noise value to large, correlated-noise modes (the weights are set to zero to completely remove a particular modes). In Fig. 4.13 we show the filtered maps using this method. noise modes which correspond to multipoles values $25 < \ell$ have been removed. These maps do not present any visible systematic feature. Only CMB and noise are expected.

We use the maps shown in Fig. 4.13 to test the distribution of the polarisation angles. We calculated the polarisation angle, $\chi = -0.5 \tan^{-1}(U/Q)$, and plotted

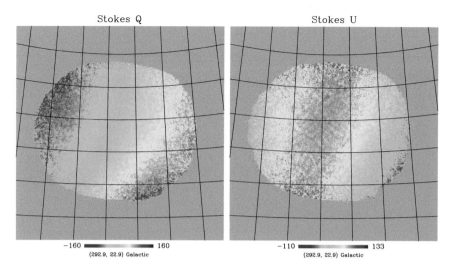

Fig. 4.10 Stokes Q and U maps of the QUIET CMB-1 field. They are dominated by large-scale modes that need to be subtracted

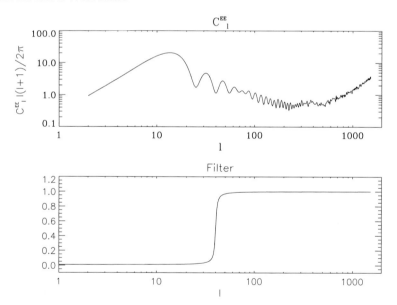

Fig. 4.11 EE power spectrum of the CMB-1 field and the filter used to remove the large scale modes

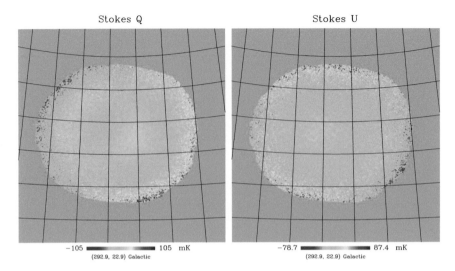

Fig. 4.12 Filtered Stokes Q and U maps of the QUIET CMB-1 field. Most of the large scale emission is removed with the use of the apodizing filter. However, there are some additional artefacts which add another spurious signal to the maps

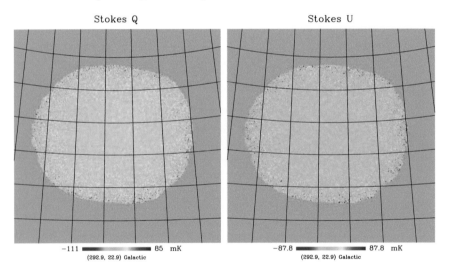

Fig. 4.13 Eigenmode filtered Stokes Q and U maps of the QUIET CMB-1 field. In this case, no visible artefacts are present in the maps. We have removed the modes equivalent to $25 < \ell$

the distribution of χ. In Fig. 4.14 we show the histogram, binned in $10°$ intervals, for all the pixels in the map.

No significant deviations from uniformity are observed. The distribution of the residuals, shown in the *bottom* panel of Fig. 4.14 is consistent with zero. This means that no particular directions are privileged in the orientation of the polarisation

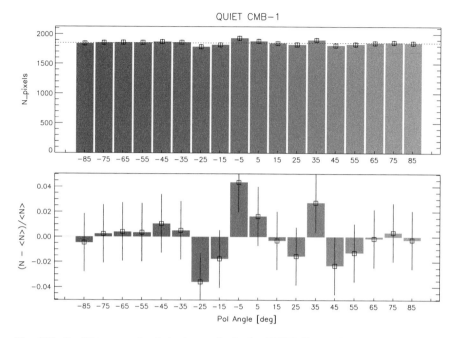

Fig. 4.14 *Top* Histogram of polarisation angles in the QUIET CMB-1 field. The *bottom panel* shows the residual histogram, after the mean value has been subtracted from all the bins. There are no significant deviations from a random distribution

vectors. No significant foreground contamination or systematic errors are likely to be present in these 95 GHz filtered maps.

4.5 QUIET CMB Results

The CMB polarisation results are published in QUIET Collaboration et al. (2011), QUIET Collaboration et al. (2012a). Here I briefly summarise the main results of these papers.

The first paper (QUIET Collaboration et al. 2011), presented the results from the Q–band array at 43 GHz. The EE and BB power spectra are measured in the multipole region $\ell = 25 - 475$. In the region corresponding to the first peak of the EE mode, namely $76 \leq \ell \leq 175$, there is a 10σ detection of EE power, confirming the measurements of the BICEP experiment (Chiang et al. 2010). The EB and BB spectra do not show any significant power and the limit obtained for the tensor-to-scalar ratio is $r < 2.1$. Foreground contamination is found in the CMB-1 field, and is consistent with polarised synchrotron emission. This was done by cross-correlating the QUIET maps with the *WMAP* 7-year 23 GHz polarisation maps. The three other fields do not show any significant contamination. These data are useful to constrain

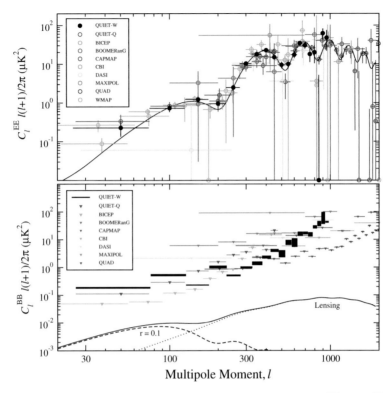

Fig. 4.15 Current observational status of the polarisation power spectra C_ℓ^{EE} and C_ℓ^{BB} as of October 2013. The EE spectra on *top* has been measured by various experiments. The upper limits on the B-modes spectrum are shown at the *bottom*. Here, the dashed line that peaks around $\ell = 80$ corresponds to the spectra produced by gravitational-waves if the tensor to scalar ratio has a value of $r = 0.1$. The second peak at $\ell \sim 1000$ is due to gravitational lensing and is not related to the primordial gravitational waves background. This figure is reproduced from QUIET Collaboration et al. (2012a)

the level of synchrotron contamination that is present in the 95 GHz data of the same fields. The QUIET Q–band maps are deeper than what *Planck* will achieve at the same frequency. The systematic effects are constrained to a level below that of any other published experiment, which corresponds to an uncertainty in the tensor-to-scalar ratio of $r < 0.03$. This was achieved by the combination of design features and observing strategy.

QUIET Collaboration et al. (2012a) presents the results of the 95 GHz observations. The EE power spectra is measured in the range $\ell = 25 - 975$, with the first three acoustic peaks observed at high signal-to-noise ratio. This spectrum is consistent with ΛCDM cosmology. The EB and BB spectra are consistent with zero, and a 95 % C.L. upper limit on the tensor-to-scalar ratio is found to be $r < 2.7$. The level of systematic uncertainty is negligible, at the $r = 0.01$ level. Some foreground contamination is detected as a correlation between the power in the CMB-1 map and the thermal-dust component, estimated using the *Planck* sky model (Delabrouille et al. 2013). The excess of power on the spectra due to this dust component is small compared to the statistical uncertainties in the power spectra. In Fig. 4.15 are shown the EE and BB power spectra measured by QUIET 43 and 95 GHz observations, as well as the measurements from various experiments. An additional conclusion of the experiment is that the use of polarisation radiometers arrays in CMB experiments provides a good way of controlling systematic errors and achieving good sensitivity.

4.6 Galactic Maps

Here we describe the QUIET Q–band Galactic maps. These maps are produced using the maximum likelihood pipeline after the careful selection of the filtering parameters. We compared them with *WMAP* Q–band polarisation maps of the same fields, showing that the QUIET maps have a better SNR. A basic analysis of the diffuse polarised emission is performed with the QUIET maps. We measure differences in the average polarisation angles between the two observed fields. These analysis are preliminary due to the still incomplete full characterisation of the noise in the QUIET maps. Nevertheless, it shows some examples of the Galactic science that can be done with these polarisation maps, which are deeper than the *WMAP* data at the same frequency.

In Fig. 4.16 we show the QUIET Q–band maps for the patch G-1, covering the Galactic centre. Stokes Q and U maps with their respective uncertainties are displayed. The central source corresponds to Sgr A, the polarised radio source at the Galactic centre. Figure 4.17 shows the polarisation maps of the G-2 field, centred at $(l, b) = [330°, 0°]$. Some artefacts remaining after data selection and filtering are still visible in the G-2 maps as near-vertical striations.

In Figs. 4.18 and 4.19 we see the comparison between the QUIET maps and the *WMAP* 9-year Q–band map, at 41 GHz, which has an angular resolution of 30′, very similar to the 28′ of the QUIET maps. The QUIET and *WMAP* maps are remarkably similar. They show a similar morphology in both Stokes Q and U maps. The *WMAP* maps are visibly noisier in the central areas of each map, where the QUIET maps have the maximum sensitivity. These maps nevertheless, are not quantitatively equivalent. The filtering of the QUIET maps to remove the atmospheric contamination and the other sources of systematics described earlier, have removed some of the extended diffuse emission. This can be appreciated in the top panel of Fig. 4.18. The *WMAP* image on the *right* shows a more extended diffuse emission along the Galactic plane.

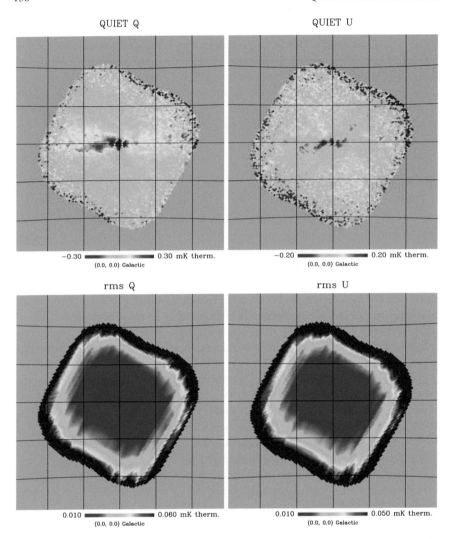

Fig. 4.16 QUIET Q–band polarisation maps of the G-1 field, centred at the Galactic centre. On the *top panel* are Stokes Q and U, and the rms noise maps for each of these are shown below. The scale is linear in all the maps and the units are mK thermodynamic. The noise maps include only thermal noise and can be interpreted as the inverse of the "number of hits" map. The grid spacing in all maps is $5° \times 5°$ in Galactic coordinates

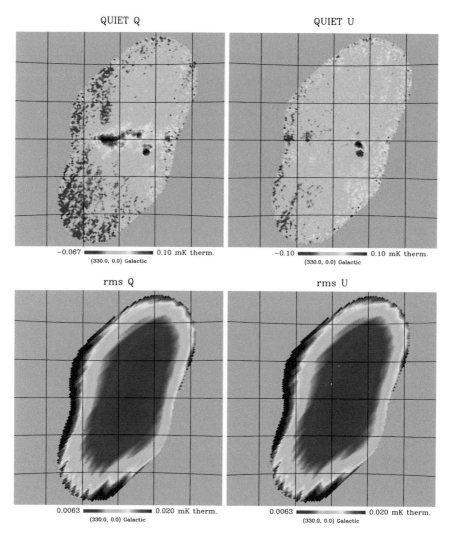

Fig. 4.17   QUIET Q–band polarisation maps of the G-2 field, at centred at $(l, b) = 330°, 0°$. On the *top panel* are Stokes Q and U, and the rms noise maps for each of these are shown below. The scale is linear in all the maps and the units are mK thermodynamic. The noise maps include only thermal noise and can be interpreted as the inverse of the "number of hits" map. The grid spacing in all maps is $5° \times 5°$ in Galactic coordinates

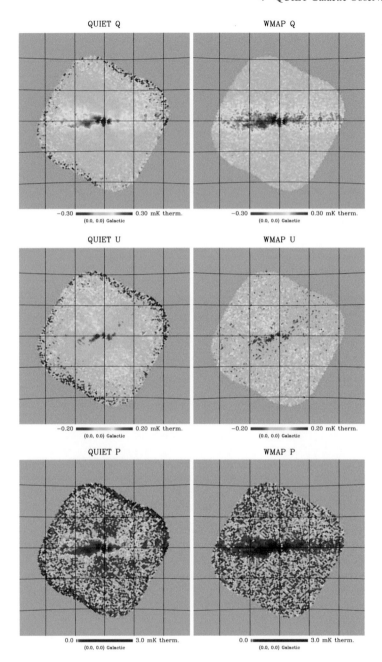

Fig. 4.18 Comparison between QUIET Q–band polarisation maps and the polarisation maps of *WMAP* Q–band for the region G-1, centred at the Galactic centre. The Stokes *Q*, *U* and the polarisation amplitude *P* are shown from *top* to *bottom*. The QUIET and *WMAP* maps are remarkably similar, showing the same overall structure. The maps from QUIET however are less noisy in the central regions. The grid in all maps is 5° × 5° in Galactic coordinates

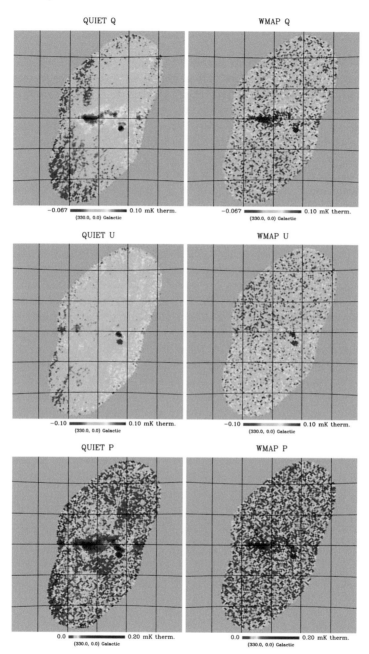

Fig. 4.19 Comparison between QUIET Q–band polarisation maps and the polarisation maps of *WMAP* Q–band for the region G-2, centred at $(l, b) = 330°, 0°$. The Stokes Q, U and the polarisation amplitude P are shown from *top* to *bottom*. The QUIET and *WMAP* maps are remarkably similar, showing the same overall structure. The maps from QUIET however are less noisy in the central regions. The grid in all maps is $5° \times 5°$ in Galactic coordinates

4.7   Spectral Index of Polarised Emission

We calculated the polarisation spectral index between QUIET and *WMAP* K-band.
Here, we used the T-T plots approach as we did in Sect. 3.3. We first converted
the maps from thermodynamic temperature to antenna temperature units, using the
formula in Eq. 2.2. We corrected for the polarisation bias using the same procedure
that we used in Chaps. 2 and 3 for *WMAP* data. Then, we smoothed all the maps to
a common 1° resolution, close to the 51′ beam of *WMAP* K–band maps to allow a
comparison with all the *WMAP* maps.

We measured the spectral indices in a strip of 6° in Galactic latitude, centred on
the Galactic plane. In Fig. 4.20 we show the T-T plots between *WMAP* K–band and
QUIET Q–band for patches G-1 and G-2. In both plots, there is a large dispersion
of the data points. This dispersion is more important for the fainter regions. For the
bright points close to the top of the correlation, the dispersion is smaller. This is most
probably due to the filtering of the QUIET maps, which affects more the diffuse
and fainter regions. The spectral indices measured in these regions are quoted in the
plots. For patch G-1, at the Galactic centre, $\beta_{K-QUIET} = -3.12 \pm 0.06$ and for patch
G-2, $\beta_{K-QUIET} = -3.12 \pm 0.07$. The error in these measurements includes a 5 %
calibration error for QUIET Q–band. These values are similar to the ones obtained
using only *WMAP* data on the plane and, due to the dispersion of the data in these T-T
plots, the uncertainties in β are similar. Using only *WMAP* data, $\beta_{K-Q} = -2.9 \pm 0.03$
for the G-1 field and $\beta_{K-Q} = -3.0 \pm 0.05$ and for patch G-2. Table 4.6 lists these
values. We note that the spectral indices measured using the QUIET maps are slightly
steeper than the ones measured using only *WMAP* data. This is probably due to the
missing signal in the QUIET data, removed by the filtering. Even though the QUIET
maps present lower noise than the *WMAP* Q–band maps, the filtering applied to
the QUIET data results in a larger dispersion of the data points in the T-T plots.
The final error in $\beta_{K-QUIET}$ is larger than the error of the spectral index measured
using only *WMAP* data. Nevertheless, both spectral indices are consistent within the
uncertainties.

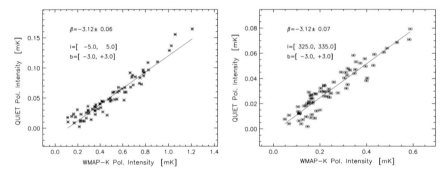

Fig. 4.20 T-T plots between QUIET 43 GHz and *WMAP* 23 GHz in the patch G-1 at the Galactic
centre (*left*) and G-2 patch (*right*), in the latitude range $-3° < b < 3°$

Table 4.6 The first column shows the polarisation spectral indices measured between *WMAP* K–band at 23 GHz and QUIET Q–band at 43 GHz, measured in the two Galactic patches G-1 and G-2

Data	$\beta_{K-QUIET_Q}$	β_{K-WMAP_Q}
Patch G-1	-3.12 ± 0.06	-2.90 ± 0.03
Patch G-2	-3.12 ± 0.07	-3.00 ± 0.05

The second column list the values measured for the same regions but this time using only *WMAP* data at *K* and *Q* bands

4.8 Polarisation Angles

In the millimetre range, polarisation is observed to be roughly perpendicular to the Galactic plane. The measurement of polarisation angles χ shows the orientation of the Galactic magnetic field. This means that the Galactic magnetic field is roughly parallel to the plane of the Galaxy. Spatial variations of χ can be used to constrain models of the Galactic magnetic field.

Figure 4.21 shows the polarisation vectors on both Galactic fields over-plotted on top of the total intensity *WMAP* Q–band map. The orientation of the vectors in the diffuse component of the maps is slightly different from patch G-1 and G-2.

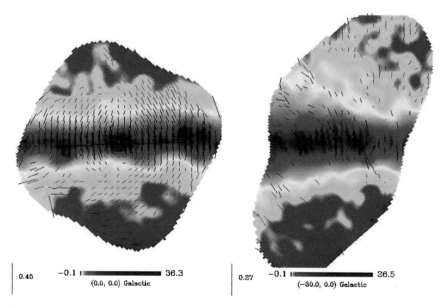

Fig. 4.21 Polarisation angles in both Galactic patches over-plotted on the *WMAP* Q–band intensity map. Pixels with signal-to-noise ratio lower than 2 in polarisation amplitude have been masked so vectors are not plotted in noisy areas. The vectors are predominantly oriented perpendicular to the Galactic plane, which means that the magnetic field is parallel to the plane of the Galaxy

Table 4.7 Mean polarization angles at different latitudes in both Galactic patches

b-range (deg)	⟨χ⟩ G-1 (deg)	⟨χ⟩ G-2 (deg)
[+3.0, +5.0]	−4.2 ± 4.0	−0.9 ± 4.4
[+1.0, +3.0]	−0.9 ± 2.1	−4.9 ± 3.9
[−1.0, +1.0]	−4.0 ± 1.3	−11.1 ± 2.5
[−3.0, −1.0]	−2.6 ± 2.7	−7.0 ± 3.5
[−5.0, −3.0]	−8.4 ± 4.2	19.2 ± 5.2
[−5.0, 5.0]	−4.0 ± 1.6	−0.6 ± 2.2

This can be appreciated by eye. On patch G-2 (*right*), the vectors closer to the Galactic plane are systematically tilted with respect to the vertical, while in patch G-1 (*left*), the distribution of angles seems more uniform.

We compared the average polarisation angles in the latitude range $b = [-3, +3]$, after masking the strong compact sources in both fields. The average polarisation angle for patch G-1 is $-1°.9 \pm 0°.4$ and in patch G-2 is $-10°.8 \pm 0°.6$, reflecting what we mentioned before. In Table 4.7 we list the average polarisation angle in different latitude cuts.

(Bierman et al. 2011), using 100, 150 and 220 GHz polarisation data, also report differences in the polarisation angle distribution across the Galactic plane. These differences reflect large scale structure in the Galactic magnetic field. It is worth pointing out that the Norma spiral arm lies in the direction of patch G-2 so is not surprising to see variations in the magnetic field towards that direction in these QUIET data.

References

Bierman, E. M., et al. (2011). A Millimeter-wave galactic plane survey with the BICEP polarimeter. *ApJ, 741*(81), 81.

Bond, J. R., Jaffe, A. H., & Knox, L. (1998). Estimating the power spectrum of the cosmic microwave background. *Physical Review D, 57*, 2117–2137.

Brown, M. L., et al. (2009). Improved measurements of the temperature and polarization of the cosmic microwave background from QUaD. *ApJ, 705*, 978–999.

Chiang, H. C., et al. (2010). Measurement of cosmic microwave background polarization power spectra from two years of BICEP data. *ApJ, 711*, 1123–1140.

Delabrouille, J., et al. (2013). The pre-launch Planck SkyModel: A model of sky emission at submillimetre to centimetre wavelengths. *A & A, 553*(A96), A96.

Dragone, C. (1978). Offset multireflector antennas with perfect pattern symmetry and polarization discrimination. *The Bell System Technical Journal, 57*, 2663–2684.

Imbriale, W. A., Gundersen, J., & Thompson, K. L. (2011). The 1.4-m telescope for the Q/U imaging experiment. *IEEE Transactions on Antennas and Propagation, 59*, 1972–1980.

Monsalve, R. A. (2010). Beam characterization for the QUIET Q–Band instrument using polarized and unpolarized astronomical sources. *Society of Photo-Optical Instrumentation Engineers (SPIE) Conference Series.* Vol. 7741. Society of Photo-Optical Instrumentation Engineers (SPIE) Conference Series.

Padin, S., et al. (2002). The cosmic background imager. *PASP, 114*, 83–97.

QUIET Collaboration et al. (2011). First season QUIET observations: Measurements of cosmic microwave background polarization power spectra at 43 GHz in the multipole range 25 ¡= l ¡= 475. *ApJ, 741*(111), 111.

QUIET Collaboration et al. (2012a). Second season QUIET observations: Measurements of the cosmic microwave background polarization power spectrum at 95 GHz. *ApJ, 760*(145), 145.

QUIET Collaboration, et al. (2012b). The QUIET Instrument. ArXiv e-prints.

Taylor, A. C., et al. (2011). The cosmic background imager 2. *MNRAS, 418*, 2720–2729.

Tegmark, M. (1997). How to make maps from cosmic microwave background data without losing information. *ApJ, 480*, L87.

Weiland, J. L., et al. (2011). Seven-year Wilkinson Microwave Anisotropy Probe (WMAP) observations: Planets and celestial calibration sources. *ApJS, 192*(19), 19.

Chapter 5
AME in LDN 1780

In this chapter we study the AME in the Lynds Dark Nebula (LDN) 1780. Here, we have detected AME using interferometric observations at 31 GHz with 6′ angular resolution (Vidal et al. 2011). Here, we studied the emission of the cloud on two angular scales. Using available ancillary data, we construct an SED between 0.408 GHz and 2997 GHz. We show that there is a significant amount of AME at these angular scales and the excess is compatible with a physical spinning dust model. In order to investigate the origin of the AME in this cloud, we use new data obtained with the Combined Array for Research in Millimeter-wave Astronomy (CARMA) that provides 2′ resolution at 31 GHz. We study the relation between the radio emission and IR radiation from dust using morphological correlations. Finally, we study the differences in radio emissivities across the cloud using a spinning dust model.

Some of the analysis presented here is based on observations using the CBI. The data reduction, imaging and preliminary analysis of the data was done during my MSc thesis. The final results, which are published in Vidal et al. (2011), were performed during the first part of my PhD.

5.1 LDN 1780 Overview

The Lynds Dark Nebula (LDN) 1780 is a high Galactic latitude ($l = 359°.0$, $b = 36°.7$) translucent region at a distance of 110 ± 10 pc (Franco 1989). It was first described in the Lynds catalogue of dark nebulae, which was created by visual inspection of red and blue optical plates (Lynds et al. 1962). Using an optical-depth map constructed from *ISO* 200 μm observations, Ridderstad et al. (2006) found a mass of $\sim 18\,M_\odot$ and reported no young stellar objects based on the absence of colour excess in point sources. In Fig. 5.1 we show a $50° \times 50°$ image of the Galaxy, marking the location of LDN 1780 with a small circle on top. A large circle also marks the Ophiuchi complex, the nearest star forming region to the Sun. LDN 1780 has a moderate column density (a few $\times 10^{21}$ cm^{-2}) that corresponds to the "translucent cloud" type

© Springer International Publishing Switzerland 2016
M. Vidal Navarro, *Diffuse Radio Foregrounds*, Springer Theses,
DOI 10.1007/978-3-319-26263-5_5

Fig. 5.1 Colour composite image showing the dust emission as seen by the *IRAS* 100 μm map and the two highest frequency channels of *Planck*, at 857 and 545 GHz. The image is shown in Galactic coordinates, with the Galactic plane running horizontally and it is about 50° on each side. The *small circle* on the *top* shows the location of LDN 1780, at $(l, b) = (359°0, 36°7)$. The *large circle* encompasses the Ophiuchi complex, the nearest star forming region to the Sun. Credit image: ESA and the HFI Consortium, IRAS

of object, i.e. interstellar clouds with some protection from the radiation field, with optical extinctions in the range $A_V \sim 1 - 4$ mag (Snow and McCall 2006).

AME has been observed in different astrophysical environments, dust clouds, HII regions, and statistically in the diffuse cirrus clouds at high Galactic latitudes (see Sect. 1.2.5 in Chap. 1). In the context of CMB foregrounds, AME originating in, or near dense dust clouds and HII regions is usually not a big concern because most of these regions lie on the Galactic plane, which is masked out from the cosmological analysis. On the other hand, AME from the diffuse cirrus needs to be treated carefully because its origin is ubiquitous across the sky. AME from cirrus clouds is difficult to study due to their shallow column density. Translucent clouds such as LDN 1780, with a column density in between that of the dark clouds and the cirrus, are therefore good targets to study AME as a CMB-foreground. Their physical properties, such as density and exposure to the interstellar radiation field (ISRF), is closer to those of the cirrus. Filling the observational gap of AME between the dense and well observed dust clouds and the diffuse cirrus is one of the motivations of this work.

LDN 1780 also has some properties that makes it a particularly good target for such studies. Ridderstad et al. (2006) found that the spatial distribution of the mid-IR emission differs significantly from the emission in the far-IR (in Fig. 5.4 this key-point can be appreciated). Also, they use IR colour ratios to show that there is an overabundance of Polycyclic aromatic hydrocarbons (PAHs) and very small grains (VSGs) with respect to the solar neighbourhood (as tabulated in Boulanger and

Perault 1988), although their result can be explained by an ISRF that is overabundant in UV photons compared to the standard ISRF (Witt et al. 2010). Because of the morphological differences in the IR, this cloud is an interesting target to make a morphological comparison with the radio data. If the AME is due to spinning dust grains, a tight correlation with IR templates is expected, in particular with the maps that trace the PAHs and VSGs. In the spinning dust model, the largest contribution comes from the smallest grains (Draine and Lazarian 1998; Ali-Haïmoud et al. 2009).

With this motivation, we have observed LDN 1780 at 31 GHz using two synthesis arrays, the Cosmic Background Imager (CBI) and new data from the Combined Array for Research in Millimeter-wave Astronomy (CARMA). In the next section we first describe the detection of AME in the cloud using CBI observations, summarising the analysis presented in Vidal et al. (2011). Then, we present the recently obtained CARMA data and also new ancillary data available, including *Planck* and *Herschel*. We use these new data to constrain the spinning dust models.

5.2 AME Detection at 31 GHz with the CBI

Here we summarise the work presented in Vidal et al. (2011). Only part of this work was done during my PhD, such as the AME emissivity analysis described in Sect. 5.2.3. We include here the descriptions that are relevant for the work with the new CARMA data, described in the next section.

We observed LDN 1780 at 31 GHz using the Cosmic Microwave Imager (CBI) (Taylor et al. 2011). The CBI was a 13-element interferometer located at 5000 m altitude in the Chilean Atacama Desert (Padin et al. 2002). It observed in ten frequency bands from 26 to 36 GHz. Each antenna has a 1.4 m dish and the primary beam is 28.′2 FWHM at 31 GHz. The angular resolution of the observations is 4.′8 FWHM. In Fig. 5.2 we show the reconstructed map of the cloud.

5.2.1 Hα Excess and Free-Free Level

The radio free-free emission must be accurately known in order to quantify the contribution of AME at GHz frequencies. One way to estimate the free-free level is using the Hα emission intensity, provided that this optical line is the result of in situ recombination and not scattering by dust.

It has been noted that LDN 1780 has an excess in Hα surface brightness. The cloud can be identified as a distinct object in the full-sky Hα map of Finkbeiner (2003). Witt et al. (2010) presents the deepest Hα map of this cloud. In Fig. 5.3 we show the continuum-corrected map from the Witt et al. (2010) work, where the diffuse cloud is well defined above the background.

There is a good correlation between the Hα and dust emission in the FIR (Burgo and Cambrésy 2006; Mattila et al. 2007; Witt et al. 2010). Burgo and Cambrésy (2006)

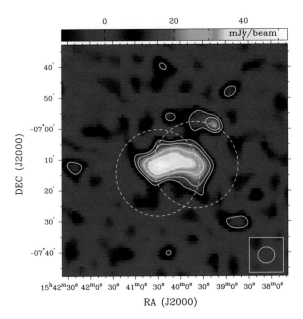

Fig. 5.2 31 GHz reconstructed image of LDN 1780, as observed by the CBI. Contours are 10, 30, 50, 70, 90 % of the peak brightness, which is 52.5 mJy beam^{-1}. The FWHM primary beam of the CBI is shown as *dashed lines* (there are two pointings). The synthesised beam is shown at the *bottom right corner*

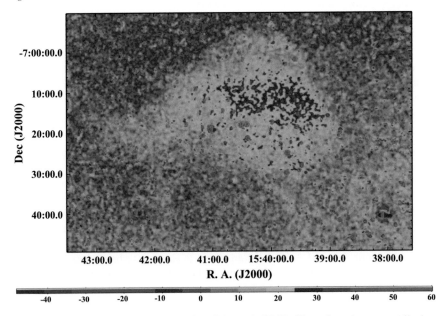

Fig. 5.3 Image of LDN 1780 Hα excess from Witt et al. (2010). Sky and continuum contributions are corrected in the narrowband Hα. The intensity scales are counts pixel^{-1}. The bulk of the diffuse emission is coincident with the central and denser regions of the cloud

found that the Hα excess is inconsistent with photo-ionisation of the cloud due to an external radiation field. This mechanism would produce a thin layer of Hα emission around the cloud, while the data shows that the bulk of the emission comes from the denser inner core of the cloud. They propose a very high rate of cosmic rays (10 times larger than the standard value) to explain the hydrogen ionisation. Mattila et al. (2007) gives a completely different explanation for the Hα excess. They suggested that it corresponds to Hα photons from the ISRF scattering from dust grains in the cloud. This has as a consequence that the Hα emission would not have a free-free counterpart as the Hα does not originate inside the cloud. Witt et al. (2010), analysing new Hα and IR data, found strong evidence favouring the scattering scenario. They showed that the observed Hα excess is fully consistent with dust scattering by typical interstellar grains in this cloud. Their results also show that the Hα excess is inconsistent with the expected surface brightness distribution produced through photoionization by external Lyman continuum photons.

Even if this is true, we can calculate the maximum free-free emission expected from the cloud under the assumption that all the Hα emission corresponds to emission from the cloud with no scattering effects. From an Hα map, a free-free brightness-temperature can be we derived. We used the relation presented in Dickinson et al. (2003), which relates the free-free radio brightness with the intensity of the Hα emission. The largest uncertainty in this conversion corresponds to the correction for dust extinction in the Hα optical map. In particular, which fraction of the total column of dust lies between the clouds and us. To estimate this effect, we take the two extreme cases to obtain limits in the free-free emission from the Hα data. First, no extinction correction, i.e. the observed Hα corresponds to all of the emission from the cloud. This map therefore translates into a lower limit for the free-free emission. In the second case, we corrected for the dust extinction using the $E(B-V)$ full-sky template from Schlegel et al. (1998) and the extinction curve given by Cardelli et al. (1989),[1] assuming that all the dust lies between the cloud and us. In this way, we produced a new Hα map which has a larger intensity than the uncorrected one, therefore producing an upper limit for the free-free emission (due to the assumption that all the dust emission present in the reddening map lies between the cloud and us, in reality, a large fraction comes from the cloud itself, therefore not absorbing some of the optical emission). Using this method, we obtained limits for the free-free emission at 31 GHz in a 1° circular aperture of $0.09^{+0.03}_{-0.01}$ Jy, where the upper and lower limits corresponds to the extinction corrected and uncorrected versions of the Hα intensity used.

---

[1] The extinction at a particular wavelength, in this case at the location of the Hα line at 656.3 nm, can be estimated from the extinction in the V band using the extinction curve. The extinction at the V band $A(V)$ is related with the *reddening* $E(B-V)$ through: $A(V)/E(B-V) = R(V)$, with $R(V)$ having a typical value of 3.1 in the Milky way.

5.2.2   IR Correlations

If dust is responsible for the 31 GHz emission in this cloud, we expect a morphological correspondence with IR emission. The 100 μm emission is due to grains bigger than 0.01 μm that are in equilibrium with the ISRF, at a temperature ∼10–20 K. On the other hand, the mid-IR emission traces VSGs at ∼100 K. These small grains are too hot to be in equilibrium with the environment. They are heated stochastically by starlight photons and, due to their small heat capacity, a single UV photon increases the particle temperature enough to emit at $\lambda < 60$ μm. The emission of these smallest grains would be proportional to their column density and also to the strength of the ISRF. This implies that, in a simple scenario, with a uniform radiation field, templates at wavelengths $\lambda < 60$ μm will trace the emission from the smallest grains, which are thought to be also the carriers of AME. Additionally, PAHs show emission line bands between 3 and 20 μm, with the strongest lines at 7.6 μm (see e.g. Draine and Li 2007 and references therein).

We compared the 31 GHz map with IR templates. In Fig. 5.4 we show the cloud in four different IR wavelengths, 12, 25, 60 and 100 μm and the contours of the 31 GHz image, as are shown in Fig. 5.2. The cloud has a different morphology in these four templates, with the peak of the emission being shifted to the West (right-hand side in the figure) with increasing wavelength. We calculated the correlation coefficient, between the CBI and IR data-sets. We found that the best correlation occurs with the template at 60 μm. This can also be appreciated by eye from the figures.

This result partially contradicts the spinning dust hypothesis where the correlation should be tighter with the NIR templates. This is a major reason for observing the cloud at higher angular resolution, which we present in Sect. 5.3.

5.2.3   AME Emissivity

An interesting result we found in this work was an anti-correlation between the hydrogen column density and the AME emissivity. Including data from additional objects observed from the same telescope we found that the AME emissivity decreases with column density. In Fig. 5.5 we show the 31 GHz emissivities plotted versus the hydrogen column density, N(H). We include a value for the cirrus clouds emissivity, estimated from the measurements at high latitudes by Leitch et al. (1997). The best-fitting power-law has a slope of -0.54 ± 0.1, and the correlation coefficient $r = -0.87$. This can be interpreted as follows. The density of very small grains (responsible for the AME in the spinning dust paradigm) decreases in denser environments as they coagulate to form larger grains. This makes the AME emissivity smaller in denser environments. This behaviour was confirmed in a larger sample of clouds observed by the *Planck* satellite (Planck Collaboration et al. 2013e).

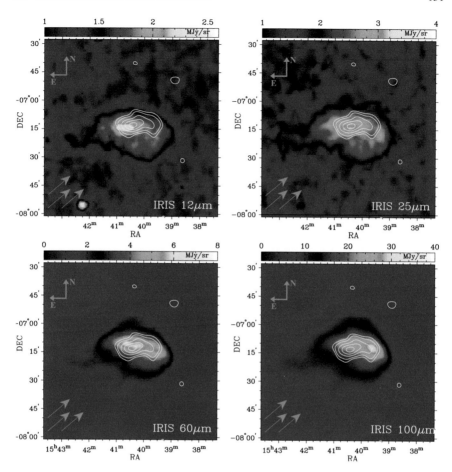

Fig. 5.4 Comparison of the restored CBI image of LDN 1780 with four IRAS templates. Contours are as in Fig. 5.2. Note the morphological differences between the IRAS bands. The *arrows* in the corner are perpendicular to the Galactic plane and point towards the north Galactic pole

The value for the slope that they find, using 98 sources is -0.43 ± 0.04, consistent with our estimation. We note however that despite the large variations in column density shown in Fig. 5.5 ($\sim 2 - 3$ orders of magnitude), the emissivities of the clouds lie in a small range of ~ 1 order of magnitude. This is good news for the interpretation of AME in different environments, as some properties measured in the denser clouds can be extrapolated to the diffuse cirrus, due to the small change of emissivity with column density detected. Vidal et al. (2011) provides more details on these results from the CBI observations.

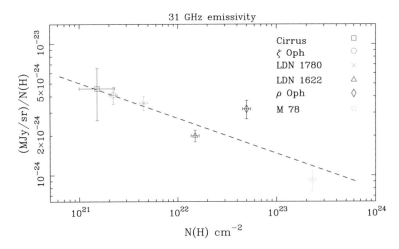

Fig. 5.5 AME emissivity at 31 GHz versus column density for different clouds observed with the CBI. The point corresponding to cirrus clouds was obtained as an average from the results of Leitch et al. (1997). The *dashed line* is the best power-law fit to the data

5.3   CARMA Data

With the aim of studying the AME from LDN 1780 at higher angular resolution than that of the CBI (≈ 4.5), we applied to obtain time in the Combined Array for Research in Millimeter-wave Astronomy (CARMA). The 3.5-m array, which consists of 8 antennas of 3.5 m diameter is well suited for observations around 30 GHz. Six "inner" telescopes are arranged in a compact configuration, with baselines ranging from 4.5 to 11.5 m. The two other telescopes provide baselines of 56 and 78 m. The receivers observe the frequency range 26–36 GHz in total intensity. The primary beam corresponds to $\approx 11'$ at 31 GHz.

This array is currently the best interferometer to detect diffuse emission in arcmin angular scales at $\sim$30 GHz. The Very Large Array or the Australia Telescope Compact Array, due to their long baselines, have a lower surface brightness sensitivity. The 3.5 m array from CARMA, has a high filling factor, allowing a better detection of diffuse emission.

We prepared a mosaic observation along the peak of the cloud, and also, to include the "gradient" of IR emission, i.e. regions where the morphology of IR differs. Originally, we prepared a 19 pointings mosaic including most of the cloud. Due to limited SNR from the first observations, we focused on three pointings, coincident with the peak of the CBI emission. In Fig. 5.6 we show the three pointing mosaic, overlaid on top of the CBI image.

In order to estimate the observation time required, we used the proportionality constant that we have measured in the cloud using *Spitzer* and CBI data between the emission at 8 μm and 31 GHz, with a value of $5.3 \pm 1.0\,\mu$K$\,$(MJy/sr)$^{-1}$ (Vidal et al. 2011). This is under the assumption that the angular frequencies of the emission are

Fig. 5.6 Mosaic pointings of the CARMA observations. The colour image shows the emission at 31 GHz as seen by the CBI. The *dashed circles* along the cloud show the location of the three pointings where we observed the cloud with CARMA, with the size of them indicating the primary beam of the CARMA antennas. At the *bottom-right corner* are shown the synthesised beam sizes for the CBI (*large*) and CARMA (*inner small*)

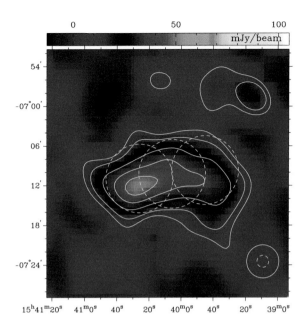

similar between the IR and 30 GHz. If we take a conservative value for the flux loss on angular scales greater than ∼12′ of 80 %, the expected brightness of the cloud close to the peak is ∼1.0 mJy/beam. We calculated the integration time required to obtain a rms noise of 0.35 mJy/beam (SNR ∼ 5) in the observed area is 25.2 h (or 4 sidereal passes) of telescope time.

5.3.1 Observations and Calibration

The observations were performed in two runs, first between 2012-06-09 and 2012-07-21 and, and between 19-05-2013 and 14-06-2013. Each run is divided into small observations blocks. In Table 5.1 we list the observation blocks. In the table we also show the "track grade", which is an internal quality flag that the observers add to the data. A grade of 100 means that the data are very good. Lower grades indicate some problem, like bad weather, a high noise temperature or that some antennas are down during the observations. If the track grade is low, the observations are repeated. This is the case for the two blocks with grade equal to 75, where no time was charged due to an antenna being disconnected during the observations.

During each one of these blocks, the source is observed along with three calibrators, which ultimately allow us to convert the measurements into astronomical flux units. Three calibrators are required to process the data. The "passband calibrator", the "flux calibrator" and the "gain calibrator" and they do not have to be necessarily different objects. The calibrators should have some ideal properties, like

Table 5.1 Observations blocks of LDN 1780

Date	Observing length [h]	Track grade
2012-06-09	2.00	100
2012-06-16	2.00	95
2012-07-03	2.60	100
2012-07-18	4.70	100
2012-07-19	3.00	100
2012-07-20	1.50	95
2012-07-21	3.00	95
2013-05-19	1.40	75
2013-05-20	3.90	75
2013-05-21	3.90	100
2013-05-28	4.25	100
2013-06-10	2.00	100
2013-06-13	5.70	100
2013-06-17	3.10	100
2013-06-14	3.30	100
Total time	46.35	

The track grade is an internal flag that the observers add to the data. A grade of 100 means very good data and a smaller grade means that the data have some problems

being constant during the observation time, an accurately known position in the sky, a relatively simple spectra and being strong enough to allow calibration in a short time.

After the observations have been made, we reduced the data and produced maps. The reduction process includes flagging, calibration and imaging. These three processes were done with the Miriad data-reduction package (Sault et al. 1995). All the tasks and routines that are mentioned in this chapter are part of this software.

We applied the following calibration steps.

- *Line length calibration.* Temperature variations can produce changes in the length of the transmission fibres. This will induce instrumental phase shifts that need to be accounted for. The routine `linecal` estimates and corrects for this effect.
- *Baseline correction.* After a reconfiguration of the array, the baseline solution will often improve during the course of that configuration. This baseline correction should be retroactively applied to the data. This is done by feeding a new antenna position file to the task `uvedit`.
- Bandpass calibration. The bandpass shape is how the receivers responds to the spectrum of the observed source. Ideally, the instrument should produce a flat spectrum if source has a flat spectral shape. In reality the receivers do not behave like this and a correction has to be made. This shape is solved by the task `mfcal`, which uses a well known spectrum of a calibrator source, in this case, the band-pass calibrator 1337-129, which is observed at the beginning and end of each observation block.

- Flux calibration. In order to convert from instrumental to flux density units, a source with a well know flux density is observed in each block. We used the 3C273 quasar for this purpose. The task `bootflux` performs this calibration.
- Phase calibration. The phases of the visibilities of the source will change during each observation block. This is due to variations in the atmosphere, system temperature or elevation of the antenna with time. A nearby source with constant flux is observed regularly (every 15 min) throughout the observation block. This allows for an interpolated solution for the phases variations during the observations. The task `selfcal` calculates and corrects the visibilities for this effect.

Even though the weather and system temperature were good during most of the time, there are some periods where some baselines are more noisy, or the atmosphere is less transparent. To avoid including these bad data into our final maps, we have to manually flag all the suspect data. This was done manually for each observation block. An example of noisy data that is flagged is shown in Fig. 5.7. In this plot, the amplitude of the phase calibrator is plotted during the observing block for each baseline. The rms of the first plot in the top-left corner, which corresponds to the baseline of the antennas 16–17, is visibly larger than the rms of the rest of the baselines. After the Line length calibration and the Baseline correction, we flagged the bad data.

In Fig. 5.8 we show the phase of the calibrator during the observations. The phase should vary smoothly with time, and this is observed in all the baselines shown in the figure. The first baseline shows a slightly larger noise. We do not include this noisy baseline in the final data.

After the phase calibration, the phases should all be centred at zero. Figure 5.9 shows the phases of the calibrator after the calibration has been done, with all of them very close to zero.

An additional sanity check after the calibration is provided by plotting the phase angle against the (u, v) distance. The longer baselines suffer atmospheric decorrelation, in which two antennas separated by a large distance see through different air masses the astronomical object. This causes the increase in the noise of the phase angle with longer baselines. This is shown in Fig. 5.10.

5.3.2 Imaging

After the visibilities are calibrated, we can produce an image. The visibilities of the three pointings are concatenated using a routine called `uvcat`. At this stage, the visibilities represent the Fourier transform of the sky, sampled in the (u, v) plane and convolved by the primary beam of the telescopes. To recover the sky image, an inversion process has to be done. The configuration of the array during the observations results in the (u, v) coverage shown in Fig. 5.11.

The visibilities are inverted using the `invert` task, which takes into account the different points and can also combine images into a mosaic. This step generates

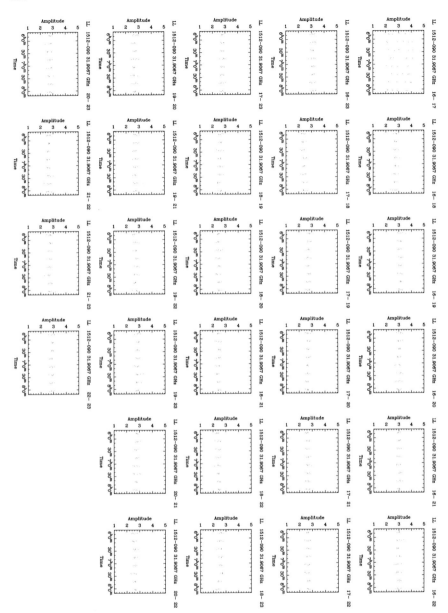

Fig. 5.7 Amplitude of the calibrator during the observation for each *baseline*. Here, the amplitude should be constant and the same for all the antennas. We have flagged the first *baseline* (16–17) due to its relatively larger noise with respect to the rest of the *baselines*

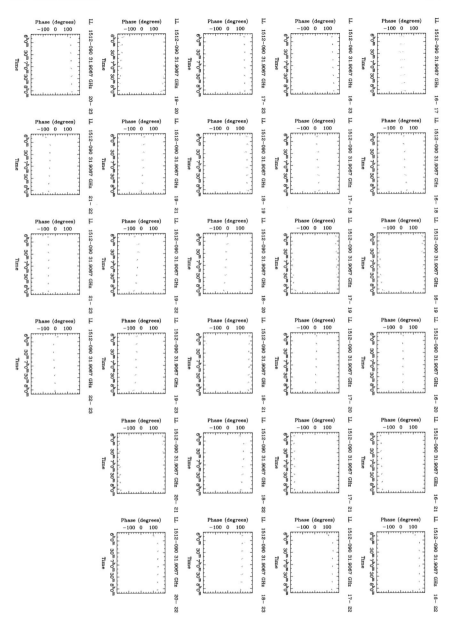

Fig. 5.8 Phase of the calibrator for each *baseline* during the observations. Ideally, these plots should have a smooth trend. When the data looks more noisy, we flag the corresponding *baseline*. In this case, we flagged the first *baseline* (16–17) as the dispersion of the visibilities is larger

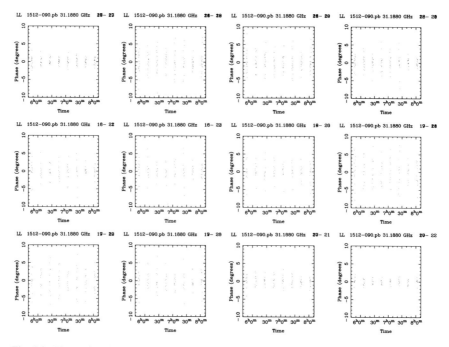

Fig. 5.9 Phase after the calibration for the phase calibrator 1512-090. All the phases are centred at zero, which is a sign of a good calibration

a "dirty map", which has many artefacts due to the incomplete sampling of the (u, v) plane. To correct for these artefacts, a deconvolution is required. There are two algorithms that are traditionally used, the CLEAN (e.g. Högbom 1974) and MEM (e.g. Cornwell and Evans 1985) reconstruction.

First we imaged the calibrator. This is used also as a test to check the calibration steps before. In Fig. 5.12 we show the dirty map of the phase calibrator 1512-090 and the deconvolved CLEAN map. The noise level in the CLEANed map is less than 1 %. It does not show visible artefacts in the linear scale that is used. We produced similar calibrator maps for each observation block in order to check the quality of the calibration.

To image the source, we tried both CLEAN and MEM reconstructions. We used natural weights in order to obtain a deeper restored image. In Fig. 5.14 we show the two maps. The synthesised beam size is plotted as an ellipse at the bottom-right corner. It has a size of $1\rlap{.}'75 \times 1\rlap{.}'45$ FWHM. Both maps present a similar morphology, but the MEM reconstruction on the right recovers more of the diffuse and extended flux.

Two radio sources from the Condon et al. (1998) catalogue are visible in the maps. They are marked with blue circles and are listed in Table 5.2.

We subtract the sources from the visibilities as we are interested in the diffuse emission from the cloud. We did the source subtraction by first using CLEAN to

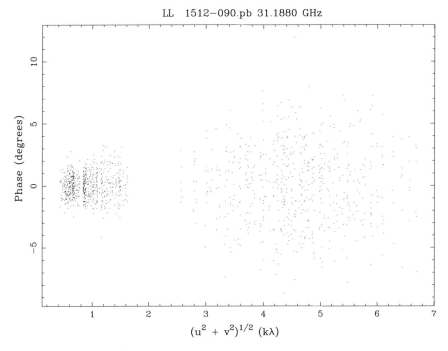

Fig. 5.10 Phase of the calibrator as a function of (u, v) distance. Due to atmospheric decorrelation, the longer *baselines* show a higher dispersion in the phase angle

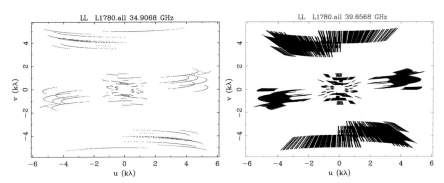

Fig. 5.11 (u, v) coverage of the CARMA array in the configuration used during the observations of LDN 1780. The plot on the *left* shows the frequency averaged coverage, while the one on the *right* shows a more filled (u, v) plane as there is no frequency averaging

obtain their flux density at the CARMA map. Then, the sources are modelled as points, located at the coordinates listed in Table 5.2 and with the flux densities that we obtained from CLEAN (also listed in Table 5.2). We visually inspected the dirty and deconvolved maps after the subtraction to check for artefacts in case of a bad

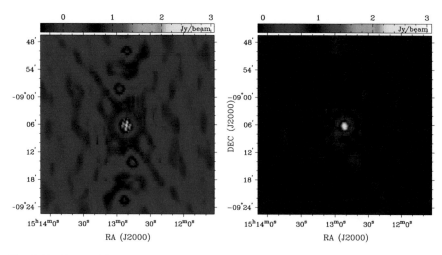

Fig. 5.12 Dirty and CLEANed map of the calibrator 1512-090. The reconstructed image on the right has a noise value which is less than 1 % of the peak value

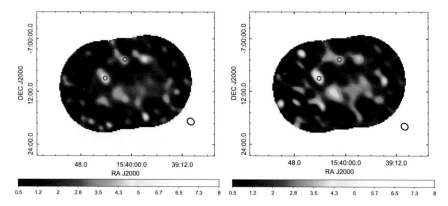

Fig. 5.13 Clean (*left*) and MEM (*right*) mosaic reconstructions of the data. The synthesised beam size is represented as a *black ellipse* at the right corner of each plot. The *two blue circles* show the position of two NVSS radio sources that lie in the area of the mosaic. The maps are plotted using a linear scale and the units are mJy beam^{-1}

estimation of the flux. Figure 5.5 shows the CLEAN and MEM reconstructions of the visibilities with the two point sources subtracted.

In order to increase the sensitivity to extended emission, we use Gaussian (u, v) tapering. This is the convolution of the visibilities with a Gaussian filter which has the effect of down-weighting the longer baselines, degrading the final angular resolution of the map and increasing the SNR for the extended emission. We used a filter size in the Fourier space equivalent to a Gaussian smoothing kernel in the image plane which produces a final map with $2'$ resolution. The rms noise in the CLEAN map is 40 % larger than the noise in the MEM map (1.4 and 0.99 mJy beam^{-1} respectively).

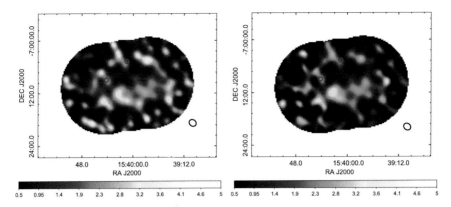

Fig. 5.14 Clean (*left*) and MEM (*right*) mosaic reconstructions of the data after the two point sources shown in Fig. 5.13 have been removed from the visibility set. The *blue circles* show the position of the sources that were subtracted

Table 5.2 Point sources that we subtracted from the visibilities, from the Condon et al. (1998) catalogue

Source	RA (J2000)	DEC (J2000)	Flux at 1.4 GHz (mJy)	Flux at 31 GHz (mJy)
NVSS J154024-070858	15:40:25	−07:08:58	13.7 ± 0.6	6.2 ± 1.3
NVSS J154006-070442	15:40:06	−07:04:43	27.1 ± 1.3	3.5 ± 1.2

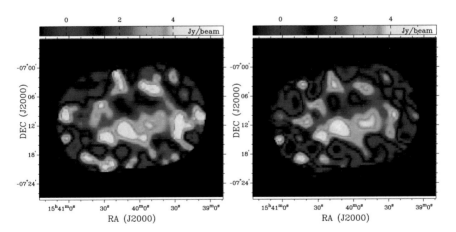

Fig. 5.15 Clean (*left*) and MEM (*right*) mosaic reconstructions of the data after the *uv* tapering, in order to increase the SNR of the extended emission. The angular resolution in this case is 2′ for both maps. The rms noise value of the CLEAN (*left*) map is 1.4 mJy beam^{-1}, and the noise of the MEM map is 0.99 mJy beam^{-1}

In Fig. 5.15 we show the two maps. We use the MEM map (*right*) for the rest of the analysis due to its smaller noise value.

5.4 Spectral Energy Distribution Over 1° Scale

We constructed the SED of LDN 1780 using 20 data points from 0.408 GHz up to 2997 GHz. Planck Collaboration et al. (2013e) performed a similar analysis on a sample of 98 clouds. Their work is based on an a semi-automatic detection and SED fitting of sources over the entire sky. Only the relatively strong sources are presented and LDN 1780 is not included in their final sample due to its low brightness. Here we show however that there is significant AME on a 1° scale in this cloud, which is consistent with spinning dust emission.

5.4.1 Data Used

We used radio and IR data to build an SED of LDN 1780 from 0.408 GHz to 2997 GHz on a 1° scale. Table 5.3 lists all the data used.

Low frequency radio data from three large surveys at 0.408, 1.42 and 2326 GHz were used to estimate the levels of synchrotron and free-free emission on LDN 1780. We used the unfiltered version of the 0.408 GHz map of Haslam et al. (1982) available at the LAMBDA website.[2] It has an angular resolution of 51′ and includes all the point sources. At 1.42 GHz the Reich et al. map (Reich 1982; Reich and Reich 1986; Reich et al. 2001) has an angular resolution of 36′. The 2.326 GHz map from Jonas et al. (1998) has an angular resolution of 20′. We assumed a 10 % uncertainty in these three data sets. An additional baseline uncertainty of 3 K (Haslam et al. 1982) is added to the 0.408 GHz map.

We included *WMAP*, which was described in Chap. 2. Here we use the five *WMAP* 9-year maps, from 23 to 94 GHz, smoothed to 1°. We assumed a conservative 4 % uncertainty (the calibration uncertainty quoted in Bennett et al. 2013, is 0.2 %). This is to account for any colour correction effect, as the spectral index of the source is not equal to the CMB spectrum.

We also include *Planck* data. *Planck* observes the full sky in nine frequency bands between 28 and 857 GHz. We used the temperature maps which were released in 2013 (DR1), and are described in Planck Collaboration et al. (2013a) and are available in the *Planck* Legacy Archive.[3]

---

[2]http://lambda.gsfc.nasa.gov/.

[3]http://www.sciops.esa.int/index.php?project=planckpage=Planck_Legacy_Archive.

Table 5.3 List of ancillary data used in the analysis

Telescope/survey	ν [GHz]	Resolution	References
Haslam	0.408	51ʹ0	Haslam et al. (1982)
Reich	1.42	35ʹ4	Reich (1982), Reich and Reich (1986), Reich et al. (2001)
Jonas	2.3	20ʹ0	Jonas et al. (1998)
WMAP 9-year	22.8	51ʹ3	Bennett et al. (2013)
Planck	28.4	32ʹ3	Planck Collaboration et al. (2013a)
WMAP 9-year	33.0	39ʹ1	Bennett et al. (2013)
WMAP 9-year	40.7	30ʹ8	Bennett et al. (2013)
Planck	44.1	27ʹ1	Planck Collaboration et al. (2013a)
WMAP 9-year	60.7	21ʹ1	Bennett et al. (2013)
Planck	70.4	13ʹ3	Planck Collaboration et al. (2013a)
WMAP 9-year	93.5	14ʹ8	Bennett et al. (2013)
Planck	100	9ʹ7	Planck Collaboration et al. (2013a)
Planck	143	7ʹ3	Planck Collaboration et al. (2013a)
Planck	217	5ʹ0	Planck Collaboration et al. (2013a)
Planck	353	4ʹ8	Planck Collaboration et al. (2013a)
Planck	545	4ʹ7	Planck Collaboration et al. (2013a)
Planck	857	4ʹ3	Planck Collaboration et al. (2013a)
COBE -DIRBE	1249	37ʹ1	Hauser (1998)
COBE -DIRBE	2141	38ʹ0	Hauser (1998)
COBE -DIRBE	2997	38ʹ6	Hauser (1998)

5.4.2 Flux Densities Measurement

To obtain the flux densities of the cloud at the different frequencies, the maps are first converted from thermodynamic to RayleighJeans (RJ) temperature units at the central frequency of each band using Eq. 2.2. Then, the maps are expressed into flux units ($Jy\,pixel^{-1}$) using the relation,

$$S = \frac{2\,k\,T_{RJ}\,\nu^2\,\Omega_{\text{pix}}}{c^2}, \tag{5.1}$$

where Ω_{pix} is the solid angle of each pixel.

The flux densities are measured by integrating the flux density over a 2° diameter aperture, centred at the position of the cloud. We subtract the background level using the median value of the pixels that lie in an annular aperture between 80′ and 100′ from the position of the source. The rms variations in this ring are used to calculate the uncertainty in the measured fluxes, including noise, CMB and background variations.

In Fig. 5.16 are displayed the 20 maps we used of LDN 1780, from 0.408 GHz up to 2997 GHz. All of them have been smoothed to a common 1° resolution. Each image is 5° on a side and the circular aperture and ring used for the photometry are indicated. The cloud is clearly visible in the high frequency maps, from 217 GHz, where the thermal dust emission dominates above the diffuse background. At lower frequencies, between 23 and 143 GHz, all the maps show a similar structure. These are due to CMB fluctuations that predominate over this frequency range and angular scale. At even lower frequencies, in the 0.408, 1.4 and 2.3 GHz maps, there is no emission from the region of the cloud visible above the background.

A main goal of the *Planck* and *WMAP* mission was to produce an accurate map of the CMB fluctuations. We use the SMICA CMB map (Planck Collaboration et al. 2013b) to subtract the CMB anisotropy from the individual frequency maps. By doing this, the LDN 1780 cloud is recognisable above the background in all the maps between 23 and 2997 GHz. Figure 5.17 show these CMB-subtracted maps. In them, LDN 1780 is easily discernible. The peak position of the source varies in some of the maps (e.g. *WMAP* 41 GHz, *Planck* 70 and *WMAP* 94 GHz). This might be due to residual emission after the CMB subtraction. We note that these maps are for illustrative purposes only since the SED fitting is done using the maps that include the CMB anisotropy.

Additional structure can be appreciated around LDN 1780 at the 40–70 GHz CMB subtracted maps in Fig. 5.17. We measured the standard deviation of these fluctuations around our cloud in the three mentioned maps, using a ring with and inner radius of 1°, centred at the location of LDN 1780, and a thickness of 3°. We compared these values for the standard deviation with the RMS noise of each map in the same aperture. The RMS noise was calculated using 500 simulations of pure noise for each map, constructed using the variance maps provided by the *WMAP* and *Planck* collaborations. Table 5.4 lists the standard deviation values around LDN 1780 in the *Planck* –44 GHz, *WMAP* –60 GHz and *Planck* –70 GHz CMB subtracted maps, as well as the RMS noise values of each of those data sets. The RMS noise values can account for at least 50 % of the measured fluctuations around LDN 1780. The additional residual fluctuations are a combination between uncertainties in the CMB map and additional foregrounds fluctuations. We measured the mean standard deviation between four different CMB maps provided by the *Planck* collaboration: SMICA, COMMANDER-RULER, SEVEN and NILC. In the same aperture, the fluctuations between these maps average a value of $2.8\,\mu K$. This value can be used as a measurement of the uncertainty of the CMB map in this region. This noise in the CMB map, in addition with the RMS noise value of the maps can account for the measured fluctuations around LDN 1780.

In Table 5.5 we list the flux densities measured using the aperture photometry. We also list the values for the flux densities measured in the CMB-subtracted map. The fluxes measured at the three lowest frequencies were negative, i.e. the background level in the ring is larger than the flux in the aperture. Because of this, we show a

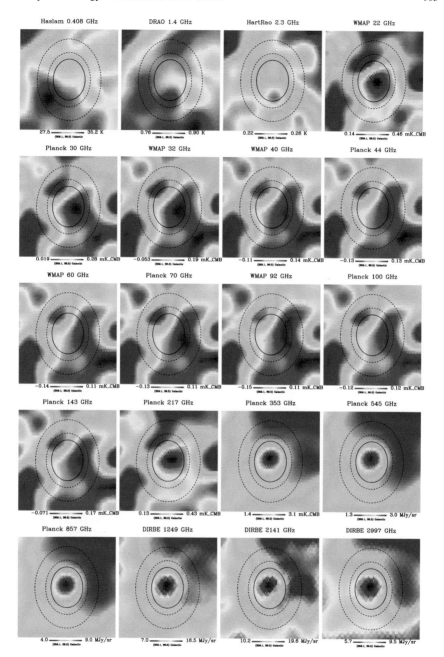

Fig. 5.16 CMB-unsubtracted maps of LDN 1780 at different frequencies ranging from the (Haslam et al. 1982) map at 0.408 GHz to the DIRBE image at 2997 GHz. The square maps have 5° in side and all have been smooth to a common angular resolution of 1° FWHM. The colour scale is linear, ranging from the minimum to the maximum of each map. The *inner circle* shows the aperture where we measured the flux and the *dashed larger circles* show the annulus used to estimate the background emission and noise around the aperture

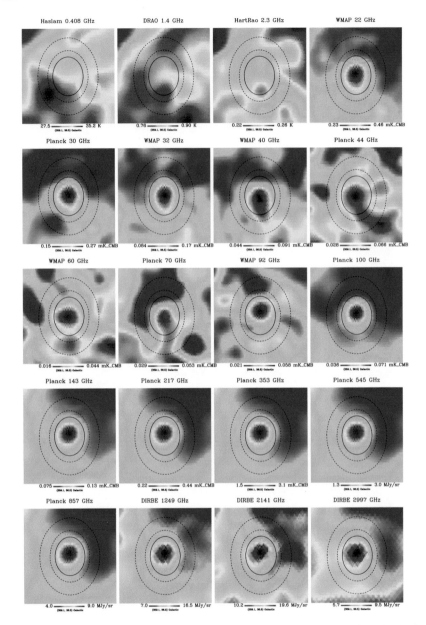

Fig. 5.17 CMB-subtracted maps of LDN 1780 at different frequencies ranging from the Haslam et al. (1982) map at 0.408 GHz to the DIRBE image at 2997 GHz. The square maps have 5° in side and all have been smooth to a common angular resolution of 1° FWHM. The colour scale is linear, ranging from the minimum to the maximum of each map. The *inner circle* shows the aperture where we measured the flux and the *dashed larger circles* show the annulus used to estimate the background emission and noise around the aperture. Compare these CMB-subtracted maps with the original ones shown in Fig. 5.16, in particularly between 23 and 217 GHz, where the CMB anisotropy dominates

Table 5.4 Standard deviation of the fluctuations visible around LDN 1780 in three of the CMB subtracted maps from Fig. 5.17

Map (GHz)	Stdev (μK)	RMS noise (μK)
Planck 44	12.0	5.9
WMAP 60	7.6	4.7
Planck 70	7.0	4.3

They are measured using a 3° thickness ring centred at the cloud, with an inner radius of 1°. The second column shows the RMS noise values measured in the same ring

Table 5.5 Flux densities of LDN 1780 over a 2° diameter aperture

Survey	Frequency [GHz]	Flux density [Jy]	CMB-sub flux density [Jy]
Haslam	0.4	<12.67	<12.67
DRAO	1.4	<0.21	<0.21
HartRao	2.3	<0.18	<0.18
WMAP	22.7	1.41 ± 0.07	1.06 ± 0.03
Planck	28.4	1.65 ± 0.12	1.12 ± 0.03
WMAP	32.9	1.34 ± 0.13	0.69 ± 0.02
WMAP	40.6	1.41 ± 0.21	0.67 ± 0.03
Planck	44	1.23 ± 0.24	0.38 ± 0.03
WMAP	60.5	1.73 ± 0.43	0.36 ± 0.04
Planck	70.4	2.47 ± 0.54	0.60 ± 0.05
WMAP	93	3.91 ± 0.91	1.08 ± 0.10
Planck	100	5.45 ± 0.95	2.17 ± 0.05
Planck	143	10.8 ± 1.5	5.28 ± 0.17
Planck	217	33.5 ± 1.8	25.7 ± 0.8
Planck	353	117 ± 3	117 ± 3
Planck	545	410 ± 13	410 ± 13
Planck	857	1189 ± 40	1189 ± 40
DIRBE	1249	1860 ± 81	1860 ± 81
DIRBE	2141	1639 ± 108	1639 ± 108
DIRBE	2997	749 ± 46	749 ± 46

3σ upper limit for each one of these points. In the next section we describe the SED fitting to the values listed in Table 5.5.

5.4.3   SED Fitting

SEDs in this frequency range are usually modelled using five components, namely synchrotron, free-free, AME, CMB and thermal-dust emission. In this case, we do not include a synchrotron component, due to the small fluxes measured at the lower frequency bands. Our model for the flux density in a $2°$ aperture is therefore described by four components,

$$S = S_{ff} + S_{AME} + S_{CMB} + S_{TD}. \tag{5.2}$$

As we discussed earlier, the free-free level in LDN 1780 itself is very low. We include a conservative upper limit for this component in the SED fitting. We use the estimation of free-free emission at 31 GHz over a $1°$ scale of $S_{31} = 0.09$ Jy shown in Sect. 5.2.1. We extrapolate this value to lower and higher frequencies using a power law with the form,

$$S_{ff} = S_{31} \, (\nu/31 \, \text{GHz})^{\beta_{ff}}, \tag{5.3}$$

where $\alpha_{ff} = -0.13$ is the free-free spectral index (for flux density) valid for the diffuse ISM (Davies et al. 2006).

The AME component is accounted for using a spinning dust model, provided by the SPDUST package (Ali-Haïmoud et al. 2009; Silsbee et al. 2011). This program calculates the emissivity in terms of the hydrogen column density j_ν of a population of spinning dust grains. It requires a number of physical parameters to generate the spectrum. We used the ideal description for the "warm neutral medium" (WNM) as defined by Draine and Lazarian (1998). The generated spectrum produces peaks at 23.6 GHz. We fit for the amplitude of this generic spectrum, so this component in the SED has only one free parameter, A_{sp}. In Sect. 5.5.2 we describe in more detail the SPDUST modelling.

A CMB component is included, using the differential form of a blackbody at $T_{\text{CMB}} = 2.726$ (Fixsen 2009). The flux density of this component has the form

$$S_{\text{CMB}} = \left(\frac{2 \, k \, \nu^2 \, \Omega}{c^2} \right) \Delta T_{\text{CMB}}, \tag{5.4}$$

where ΔT_{CMB} is the CMB anisotropy temperature, in thermodynamics units.

The dust emission at wave lengths $\lambda > 60 \, \mu$m is usually described using a modified blackbody model. The flux density measured in a solid angle Ω is,

$$S_{TD} = 2 \, h \, \frac{\nu^3}{c^2} \frac{1}{e^{h\nu/kT_d} - 1} \, \tau_{250} (\nu/1.2 \, \text{THz})^{\beta_d} \, \Omega, \tag{5.5}$$

where k, c and h are the Boltzmann constant, the speed of light and the Planck constant respectively; T_d is the dust temperature and τ_{250} is the optical depth at $250 \, \mu$m.

We used the MPFIT IDL package (Markwardt 2009) to calculate the least-squares fit. There are two *Planck* bands, centred at 100 GHz and at 217 GHz, which can include significant amount of CO line emission (Planck Collaboration et al. 2013c), corresponding to the transitions $J = 1 \rightarrow 0$ at 115 GHz and $J = 2 \rightarrow 1$ at 230 GHz. LDN 1780 is know to have a molecular component (Laureijs et al. 1995). To avoid contamination from these lines in the fluxes measured at these bands, we did not

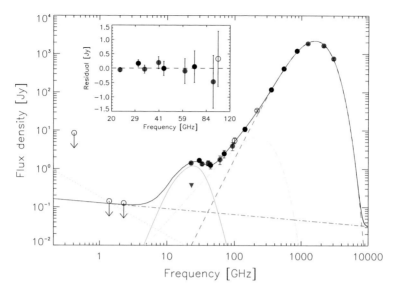

Fig. 5.18 Spectra of LDN 1780 including low frequency data, *WMAP* (*blue dots*), *Planck* (*black dots*) and *COBE* -DIRBE data (*red dots*). The *black line* is the best fit to the data, which includes four components. Thermal dust emission is represented with a *red dashed line*. A CMB component is shown in *yellow*. The *blue line* represent an upper limit for the free-free emission as expected from the Hα map. The *blue triangle* shows the free-free emission expected from the *WMAP* 9-year MEM template. The *green line* represents the spinning dust model. The two points that can be contaminated by CO emission line, at 100 GHz and at 217 GHz, are shown as a empty *black circle*. The *cyan dotted line* illustrates the slope of a synchrotron spectral index ($\beta = -3.0$). The fit do not include a synchrotron component. The insert shows the residuals (data–model) around the region where the spinning dust component is important. The residuals are consistent with zero

include these two channels in our fit. In Fig. 5.18 we show the best fit to the data. Figure 5.19 shows the best fit to the CMB-subtracted data. In both figures, the low frequency data are represented with 3σ upper limits. The largest uncertainty at 0.408 GHz comes from the ± 3 K absolute baseline uncertainty level, this large uncertainty is most probably overestimated. The blue triangle at 23 GHz represents the expected free-free level predicted by the *WMAP* MEM method (discussed in Sect. 3.5). No significant CO contribution at 100 GHz and at 217 GHz is obvious in the SED. We remind that this MEM map over-predicts the free-free emission in this cloud as the observed Hα intensity corresponds to scattered photons from the ISRF. Finally, in order to illustrate the spectrum shape of a synchrotron component, we include a dotted line fixed to the upper limit at 1.4 GHz, with a synchrotron slope ($\alpha = -1$). We do not include a synchrotron component in the fit. In Table 5.6 we list the parameters and uncertainties derived from the fit.

The difference in the fit parameters between the CMB-subtracted data and the "normal" maps is small and consistent with zero within the uncertainties. The CMB component fitted in the CMB-subtracted maps is consistent with zero (we did not fix the ΔT_{CMB} amplitude to zero in this fit to check for any residual CMB component).

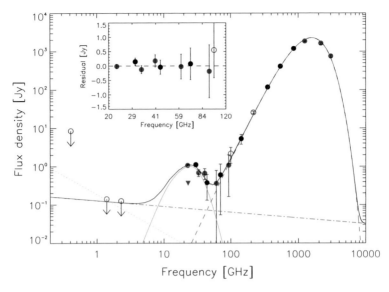

Fig. 5.19 Same as Fig. 5.18, but this time, using the CMB-subtracted maps

Table 5.6 Fitted parameters for the SED of LDN 1780

Type	$\tau_{250}[\times 10^{-5}]$	$T_d[K]$	β_d	$A_{sd}[10^{20}\,cm^{-2}]$	ΔT_{CMB}	$\chi_r^2\,[\mu K]$
Normal	2.4 ± 0.3	16.6 ± 0.5	1.62 ± 0.09	3.0 ± 0.2	14.7 ± 2.0	1.0
No–CMB	2.9 ± 0.3	16.0 ± 0.3	1.77 ± 0.05	2.6 ± 0.1	0.5 ± 0.4	2.6

Also listed is the reduced χ^2 of the fit. The second column lists the parameters of the fit using the CMB-subtracted maps

The fact that the measured $\Delta T_{CMB} = 0.5 \pm 0.4\,\mu K$ shows consistency within the two fits. The χ_r^2 is worse when using the CMB-subtracted maps. A reason for this is that the error bars of the data points in this case are smaller, because the fluctuations in the annular aperture are much smaller after subtracting the CMB anisotropy. In this cloud, the CMB contribution is significant, but we show that it is well quantified.

These SEDs show that there is AME present in this cloud. If we fit to zero the spinning dust component of the fit, the overall fit is very poor, giving a $\chi_r^2 = 9.3$.

5.5 Dust Properties at 2′ Resolution

Using the IR data available at angular resolutions similar to our CARMA maps of 2′, we can obtain some physical properties of the cloud, such as its temperature and column density. We do this by fitting for the spectrum of the thermal dust emission (as defined in Eq. 5.5) in each pixel. For this fit, we use five data points, at 70, 160, 250, 350, and 500 μm, which are dominated by thermal dust. In Fig. 5.20 we show these

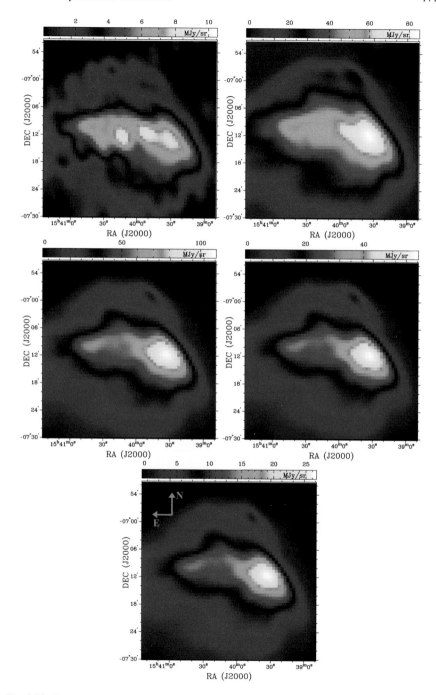

Fig. 5.20 IR maps of LDN 1780. On *top* are the 70 μm (*left*) and 160 μm (*right*), in the *middle* are 250 and 350 μm and at the *bottom* is the map at 500 μm. The map at 70 μm shows a morphology different to all the other maps. The peak here is shifted to the East (*left*)

five IR maps. All the FIR maps (from 160 to 500 μm) show a similar morphology. The map at 70 μm is slightly different to the rest, with the peak of the emission shifted towards East (left in the figure). This is due to the contribution at this band of emission from very small grains (VSGs), which are hotter towards the East due to the larger intensity of the radiation field in this direction.

Due to the smaller number of data points that we have here in comparison to the previous fit at 1° scales (5 vs. 20), we fix the dust spectral index to $\beta_d = 1.8$, a value similar to the one measured in the 1° fit. This means that in this case our modified black-body fit only has two parameters, the optical depth τ_{250} and the temperature of the big grains. We have calculated the fit only where the signal-to-noise ratio of the pixels is larger than 2. In Fig. 5.21 we show the resulting map for the optical depth at 250 μm and for the temperature of the dust. The colder regions corresponds to where the optical depth is larger. This is expected as these regions are more shielded from the ISRF.

From the optical depth map, we can obtain a hydrogen column density map using the linear relation $\tau_{250}/N_H = 2.32 \pm 0.3 \times 10^{-25}\,\text{cm}^2$, as measured by Planck Collaboration et al. (2011). From the temperature map, we can derive a map for the radiation field. The radiation field G_0 can be estimated using the following relation (Ysard et al. 2010),

$$G_0 = \left(\frac{T_d}{17.5\,[\text{K}]}\right)^{\beta_d+4}, \qquad (5.6)$$

where the spectral distribution of the radiation field is assumed to have the standard shape, defined in (Mathis et al. 1983). Planck Collaboration et al. (2013d) has

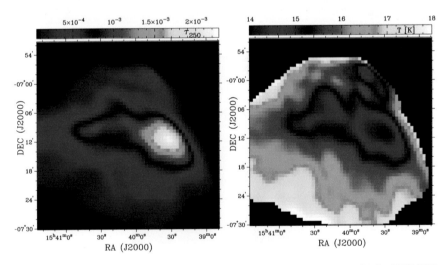

Fig. 5.21 Map of optical depth at 250 μm (*left*) and the dust temperature (*right*) for LDN 1780. The coldest areas of the cloud, at $T_d \approx 14\,\text{K}$ correspond to regions with larger optical depth. This is expected as the denser regions are more shielded from the radiation field

recently shown that this relation might not hold in every environment, and variations in T_d might be due to variations in dust properties, such as grain structure or size distribution. We will use this map in the following section.

5.5.1 IR Correlations

Here we investigate correeelations with the IR data. We include the *Spitzer* IRAC map at 8 μm, which traces PAHs, as well as the *Spitzer* MIPS map at 24 μm, tracing VSGs. To calculate the spatial correlations, we selected a rectangular region of 25′ × 15′ around the centre of the CARMA mosaic. We smoothed all the maps to a common 2′ resolution, the same as the CARMA map at 31 GHz. We used Spearman's rank correlation coefficient, r_s, to quantify the correlation of the two maps in a pixel-by-pixel comparison. This coefficient has the advantage, over the traditional Pearson correlation coefficient, that the relation between the two variables that are being compared does not have to be linear. A Spearman's rank of $r_s = 1$ will occur when the two quantities are monotonically related, even is this relation is not linear. The uncertainties in r_s are estimated using 1000 Monte Carlo simulations calculated using the uncertainties in the maps.

We calculated r_s between the 31 GHz map and the IR templates at 8, 24, 70, 160, 250, 350 and 500 μm. Because the IR emission of the smallest grains depends on the radiation field, we also calculated the correlation of the 31 GHz data with the IR templates divided by the radiation field map ($G_0 \propto T_d^6$) to account for the differences in the IRF across the cloud. The maps that are corrected by G_0 should be better tracers of the column density of small grains than the original maps. In Table 5.7 we list the values of r_s for the different IR templates for both the original IR maps and also the versions corrected by G_0. Among the original maps, the best correlation is between the 31 GHz map and 70 μm template. The NIR maps at 8 and 24 μm, present the lowest correlation coefficient, and the maps at 160, 250, 350 and

Table 5.7 Spearman's rank between the 31 GHz map and the different IR templates

Wavelength [μm]	r_s	r_s after G_0 correction
8	0.14 ± 0.06	0.38 ± 0.07
24	0.21 ± 0.06	0.46 ± 0.06
70	0.49 ± 0.07	0.45 ± 0.07
160	0.36 ± 0.07	0.31 ± 0.07
250	0.35 ± 0.06	0.31 ± 0.07
350	0.34 ± 0.06	0.30 ± 0.07
500	0.34 ± 0.06	0.30 ± 0.06

The column on the right shows the correlation value for the IR maps after they have been divided by the radiation field map, in order to account for variations in the illumination of the grains across the cloud

$500 \,\mu$m show similar r_s as expected, as these maps are tracing the same population of large grains. After dividing by the G_0, the correlation with the NIR maps at 8 and $24 \,\mu$m improves by a factor larger than 2. This increase is significant and can be appreciated even by eye. In Fig. 5.22 we show on *top* the original 8, 24, 70 and $160 \,\mu$m maps of the cloud. At the *bottom*, we show the same maps after being divided by the G_0 map. The G_0-corrected 8 and $24 \,\mu$m maps present a morphology closer to the 31 GHz black contours. The morphology of the 70 and $160 \,\mu$m does not change as much after the correction.

The peak of the 31 GHz emission in LDN 1780 is not coincident with the regions with higher column density.

5.5.2   Spinning Dust Modelling

The peak of the 31 GHz emission in LDN 1780 is not coincident with the regions with higher column density. This implies a larger radio emissivity from the less dense regions. Here we investigate if such variations in emissivity can be explained using a spinning dust model.

We compare the emissivity at the peak of the 31 GHz map, with the emissivity at the region with largest column density of the cloud. In Fig. 5.23 we show these two regions and in Table 5.8 we list average values over a $2'$ diameter aperture for the column density, dust temperature, and the relative intensity of the ISRF obtained from the maps produced in Sect. 5.5. We also list the flux densities at 31 GHz in the $2'$ aperture and the ratio of the flux with the mean hydrogen column density.

Region No. 1 shows a 31 GHz emissivity which is $18.7/3.2 = 5.8$ times larger than that of Region No. 2. We will see if the the SPDUST package from Ali-Haïmoud et al. (2009), Silsbee et al. (2011) can produce emissivities different by a factor $\sim$5.8 within this cloud.

In SPDUST, there are seven input parameters that are related to the environmental conditions of the emitting region. These are,

1. Total hydrogen number density n_H.
2. Gas temperature.
3. Intensity of the radiation field relative to the average interstellar radiation field G_0.
4. Hydrogen ionization fraction $x_H \equiv n_{H+}/n_H$.
5. Ionized carbon fractional abundance $x_C \equiv n_C + /n_H$.
6. Molecular hydrogen fractional abundance $y \equiv 2n(H_2)/n_H$.
7. Parameters for the grain size distribution: a set of dust parameters taken from a Table in Weingartner and Draine (2001).

Some of these parameters have a significant effect on the amplitude and peak frequency of the spinning dust spectrum that they produce. In order to visualise the effect, we have created spectra for a range of parameters. We first set all the parameters to the properties of a molecular cloud (which are expected to be similar

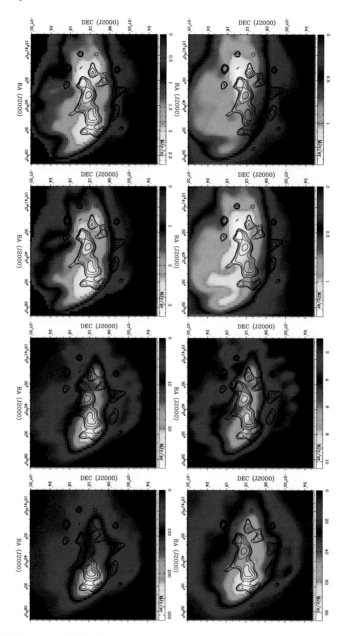

Fig. 5.22 NIR maps of LDN 1780. The original maps are on the *top* and those corrected by the radiation field G_0 are on the *bottom*. From *left* to *right*, the first column is the 8 μm, the second 24 μm, the third 70 μm and the fourth 160 μm. The contours of the 31 GHz CARMA map are overlaid on all the maps. The G_0-corrected 8 and 24 μm maps (the two *bottom left* maps) show a better correlation with the 31 GHz data than the original 8 and 24 μm maps. The opposite occurs for the 70 and 160 μm (the two *bottom right* maps). In this case, the correction for the ISRF results in a worst correlation with the 31 GHz map compared to the original. The quantitative values of the correlation are listed in Table 5.7

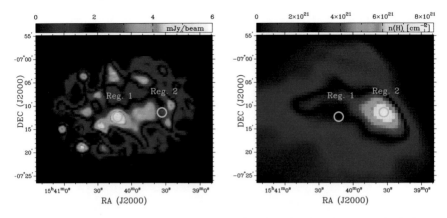

Fig. 5.23 On the *left* is the 31 GHz CARMA map of LDN 1780 and on the *right* is the hydrogen column density map obtained from the IR data in Sect. 5.5. Region 1 corresponds to the peak of the CARMA map while region 2 in centred at the peak of the column density map on the *right*. Both circular regions have $2'$ diameter

Table 5.8 Parameters of the regions shown in Fig. 5.23

Region	$N(H)$ [$\times 10^{21}$ cm^{-2}]	T_d [K]	G_0 (mJy)	$S_{31} \times 10^{-24}$ [Jy cm^{-2}]	$S_{31}/N(H)$
1	2.4	16.6	0.7	4.5	18.7
2	7.3	15.0	0.4	2.4	3.2

The column density, dust temperature and radiation field intensity G_0 are obtained from the map produced in Sect. 5.5. The fluxes at 31 GHz are integrated over a $2'$ diameter aperture. Their location is shown in Fig. 5.23

Table 5.9 SPDUST parameters of the regions shown in Fig. 5.23

Region	$n(H)$ cm^{-3}	T [K]	G_0	x_H	x_C	y	Line from WD2001
MC	300	20	0.01	0	0.0001	0.99	7
CNM	30	100	1	0.0012	0.0003	0	7
Reg 1	1000	56	0.7	0.0012	0.0003	0.5	7
Reg 2	4000	40	0.4	0	0.0001	0.99	7

The column density, dust temperature and radiation field intensity G_0 are obtained from the map produced in Sect. 5.5. The flux densities at 31 GHz are integrated over a $2'$ diameter aperture. Their location is shown in Fig. 5.23

to the LDN 1780 conditions), as defined in Draine and Lazarian (1998). In Table 5.9 we list these values. Then, we modified each parameter within a range of values, while leaving the rest fixed, and plotted the resulting spectra. In Fig. 5.24 we show the resulting spectra. First is the change in the spectrum with respect to hydrogen density. In general, the emissivity and peak frequency increase with density. A similar effect occurs for a changing temperature, radiation field and ionisation fraction; there is an increase in amplitude and peak frequency of the spectra proportional to these parameters. The variation in molecular gas fraction, y, does not alter the resulting

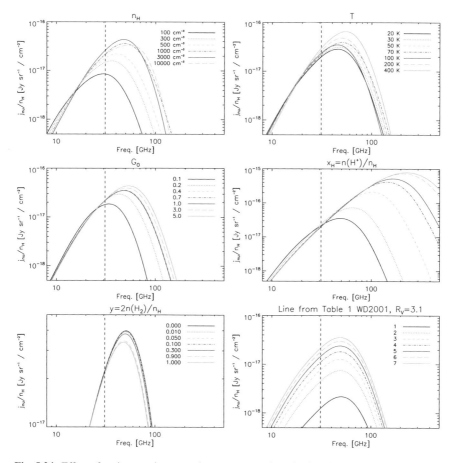

Fig. 5.24 Effect of various environmental parameters to the spinning dust spectra. First we have fix all the parameters to the MC environment, as defined in Table 5.9, and then we have varied individually each parameter. We show, from *top-left* to *bottom-right* the effect produced by varying: the hydrogen density n_H, the gas temperature T, the intensity of the radiation field G_0, the hydrogen ionisation fraction x_H, the fraction of molecular gas y, and the parameters that define the grain size distribution, as defined in the different lines of Table 5.1 from Weingartner and Draine (2001). The vertical *dashed line* in all the plots marks the frequency of our CARMA data at 31 GHz

spectra considerably in this environment. The grain properties, characterised by the parameters listed in Table 5.1 from Weingartner and Draine (2001) produce a large variation in the amplitude of the spectra, but not in the peak frequency.

As we can see from the figure, the different parameters can yield amplitudes and peak frequencies that can be very different. In order to constrain the parameter space, we use information about the physical properties of LDN 1780 known from previous studies. Mattila and Sandell (1979) observing neutral hydrogen and OH with the 100-m Effelsberg radio telescope found that the kinetic temperature of hydrogen was in the range $T_k = 40 - 56$ K. They also quote a mean value for the total density of the gas of $n = 1.8$ cm^{-3}. Laureijs et al. (1995) finds an average total density of 10^3 cm^{-3}, while Toth et al. (1995) obtains a lower value of 0.6×10^3 cm^{-3}. These two works also show that the cloud is in virial equilibrium and presents an r^{-2} density profile.

We can assume that Region No. 1, which corresponds to the peak of the 31 GHz map, has a density equal to the average density of the cloud of 1000 cm^{-3}. This is reasonable as the column density at that point has a near-average value over the cloud (see left panel of Fig. 5.23). As this point is close to half-way to the border of the cloud, and we know that the density profile here presents an r^{-2} dependency, we can estimate the value of the highest density of the cloud to be $0.5^{-2} = 4$ times larger than the average. We can also assume that the coldest region of the cloud will be the one with higher column density (this region also has the lowest dust temperature, as we shown in Sect. 5.5). We therefore assign to Region 2, the lowest gas temperature allowed by the work of Mattila and Sandell (1979), $T = 40$ K. The largest temperature that Mattila and Sandell (1979) predicts for the cloud, $T = 56$ K is assigned to Region 1. Probably the real temperature of Region 1 is lower than this value (and therefore closer to the temperature of Region 2) as the Region 1 is not at the border of the cloud, where the exposure to UV photons is larger. We use the values for the radiation field that we have calculated for both regions. Region 2, at the peak of the column density, is also coincident with a peak in the ^{13}CO map from Toth et al. (1995), this implies that the ionisation fraction in this region is very close to zero, due to the molecular nature of the gas at this position. For the ionisation fraction, we use the value that Draine and Lazarian (1998) defines for the cold neutral medium (CNM), as listed in Table 5.9. The carbon ionisation fractions for the two regions are also taken from the conditions for MC and CNM in Draine and Lazarian (1998). We list the parameters for Regions 1 and 2 in Table 5.9. *We note that the absolute value of these parameters is not as important as the ratio between them, as we want to compare the ratio of the emissivities produced by* SPDUST.

In Fig. 5.25 we show the resulting spectra for regions 1 and 2, using the parameters listed in Table 5.9. The difference in the parameters produces a difference in emissivity that is almost zero at 31 GHz and only 21 % at the peak of each spectrum, around 75 GHz. This is not strange due to the small variation in the parameters from region 1 to region 2. The variations previously shown in Fig. 5.24 are important when the parameters are changed by a larger fraction than the ones we have used here. This shows that variation in the environmental conditions are not sufficient to explain the emissivity differences observed in the cloud.

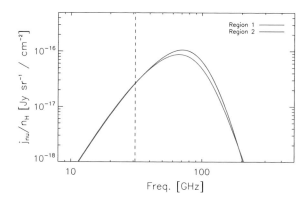

Fig. 5.25 SPDUST spectra for regions 1 and 2, using the parameters listed in Table 5.9. The *vertical dashed line* in all the plots marks the frequency of our CARMA data, 31 GHz

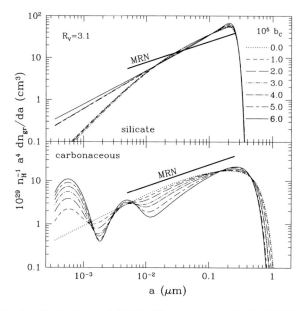

Fig. 5.26 Grain size distribution modelled by Weingartner and Draine (2001). The *top panel* shows the distribution for silicate dust, which shows a power-law behaviour for the smallest grains towards the left. The *bottom panel* shows the carbonaceous dust, where two "bumps" at the low-size end represent the smallest PAHs, with up to ~ 1000 atoms. The different *dashed lines* show the distribution for a different abundance of C, b_C. For a larger amount of carbon, the fraction of PAHs will increase. This figure is taken from Weingartner and Draine (2001)

Variations in the grain properties however could account for the differences. The differences in the IR morphology of the cloud can be explained by a difference in the type of grains across the cloud. Table 5.1 from Weingartner and Draine (2001) lists values for the parameters that define a functional form for the grain size distribution. In Fig. 5.26 we show an example of different grain size distributions from that paper.

The silicate grains show a power-law shape for the smallest grains, which are the ones responsible for most of the spinning dust emission. The carbonaceous grains on the other hand, show a more complicated distribution, with two "bumps" in the small-size regime. These local peaks are the small PAHs, large molecules with less than $\sim 10^3$ atoms. In the figure, the different dashed lines that are plotted in Fig. 5.26, represent different values of b_C, the total C abundance per H atom. A larger abundance of C, increases the proportion of PAHs.

The number of PAHs in the SPDUST model, which is characterised by the relative size of the bumps in the grains size distribution, produces large differences in the emissivity at radio frequencies. If we allow this value to change across the cloud, we can reproduce the observed differences in radio emissivity. In Fig. 5.27 we show the SPDUST spectra for region 1 and 2 using the same parameters of Table 5.9, but this time changing the "line" parameter to represent a variation in carbon abundance of a factor 6. By doing this, the emissivity ratio at 31 GHz is 5.5, close to the measured 5.8 from the CARMA data.

A factor of six in the total carbon abundance is difficult to explain giving the proximity of the two regions in the same cloud. An alternative to such a drastic change in the chemical composition of the cloud is to modify the relative size of the PAH bumps in the grain size distribution, for a fixed total carbon abundance. The values quoted by Weingartner and Draine (2001), of 0.75 and 0.25 for the amplitude of the large and small peaks seen in the bottom panel of Fig. 5.26 represent a best-fit value to a number of different clouds, so they will likely differ from cloud to cloud. We have modified the SPDUST package to allow the modifications of these parameters. In Fig. 5.28 we show the spinning dust emissivities for the MC parameters, varying the relative proportion of the amplitude of the two PAH "bumps" from Fig. 5.26. The different curves represent different values for the amplitude of the first (smallest grains) peak b_{C1}. It is clear that variations in these parameters can change the spinning dust emissivity by a large factor (up to ~ 3 orders of magnitude).

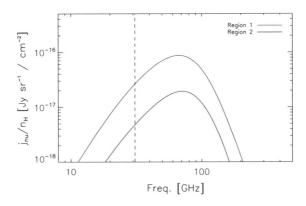

Fig. 5.27 SPDUST spectra for regions 1 and 2, using the parameters listed in Table 5.9, but this time, changing the "line" parameter, to change the grain size distribution by increasing the total carbon abundance. In this case, the ratio of the two spectra at 31 GHz (showed with the *dashed line*) is 5.5, close to the measured 5.8 from the CARMA data

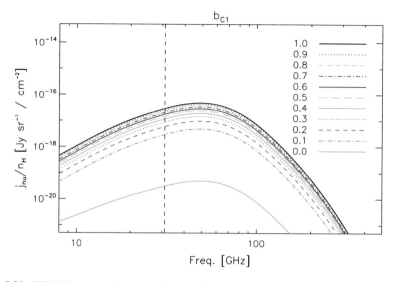

Fig. 5.28 SPDUST spectra for the conditions of the molecular cloud environment as defined in Table 5.9, but changing the relative amplitude of the two PAH bumps from the grain size distribution. The total abundance of carbon is fixed

We run SPDUST to find the relative amplitude of the PAH bumps that can reproduce the observed 5.8 factor in emissivity. We found that in region 1, the amplitude of the first PAH peak in the grain size distribution has to be 9 times larger than in region 2 to obtain a ratio in the emissivities close to the observed one. Figure 5.29 shows the resulting spectra for both regions.

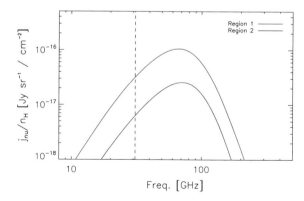

Fig. 5.29 SPDUST spectra for regions 1 and 2, using the parameters in the grain size distribution listed in Table 5.9, but this time, changing the relative amplitude of the PAH bumps, for a fixed total carbon abundance. In this case, the ratio of the two spectra at 31 GHz (shown with the *dashed line*) is 5.1, close to the measured 5.8 from the CARMA data

It is clear that, giving the degrees of freedom of SPDUST, the observed differences in emissivity across the cloud are compatible with spinning dust. Nevertheless, we have shown that if we constrain the parameter space to be compatible with the physical properties of the cloud found in the literature, the variations in 31 GHz emissivity observed in the CARMA data cannot be explained only by environmental variations. A difference in the grain properties in the two regions in necessary. These differences are expected, as there are differences in the IR morphology of the cloud. In this scenario, the denser region of the cloud, which shows low radio emission has a smaller proportion of the smallest PAHs, and this fraction is $\sim$9 times larger at the peak of the 31 GHz emission. The origin for this difference might be related to the coagulation of the smallest grains in the denser and colder region of LDN 1780.

References

Ali-Haïmoud, Y., Hirata, C. M., & Dickinson, C. (2009). A refined model for spinning dust radiation. *MNRAS, 395,* 1055–1078.

Bennett, C. L., et al. (2013). Nine-year Wilkinson Microwave Anisotropy Probe (WMAP) Observations: Final Maps and Results. *ApJS, 208*(20), 20.

Boulanger, F., & Perault, M. (1988). Diffuse infrared emission from the galaxy. *I - Solar neighborhood. ApJ, 330,* 964–985.

Cardelli, J. A., Clayton, G. C., & Mathis, J. S. (1989). The relationship between infrared, optical, and ultraviolet extinction. *ApJ, 345,* 245–256.

Condon, J. J., et al. (1998). *The NRAO VLA Sky Survey. AJ, 115,* 1693–1716.

Cornwell, T. J., & Evans, K. F. (1985). A simple maximum entropy deconvolution algorithm. *A & A, 143,* 77–83.

Davies, R. D., et al. (2006). A determination of the spectra of Galactic components observed by theWilkinsonMicrowave Anisotropy Probe. *MNRAS, 370,* 1125–1139.

del Burgo, C., & Cambrésy, L. (2006). Properties of dust and detection of Hα emission in LDN 1780. *MNRAS, 368,* 1463–1478.

Dickinson, C., Davies, R. D., & Davis, R. J. (2003). Towards a free-free template for CMB foregrounds. *MNRAS, 341,* 369–384.

Draine, B. T., & Lazarian, A. (1998). Electric dipole radiation from spinning dust grains. *ApJ, 508,* 157–179.

Draine, B. T., & Li, A. (2007). Infrared emission from interstellar dust. IV. The Silicate-Graphite-PAH Model in the Post-Spitzer Era. *ApJ, 657,* 810–837.

Finkbeiner, D. P. (2003). A Full-Sky Ha template for microwave foreground prediction. *ApJS, 146,* 407–415.

Fixsen, D. J. (2009). The temperature of the cosmic microwave background. *ApJ, 707,* 916–920.

Franco, G. A. P. (1989). High latitude molecular clouds—distances derived from accurate photometry. *A & A, 223,* 313–320.

Haslam, C. G. T. et al. (1982). A 408 MHz all-sky continuum survey. II—the atlas of contour maps. *A & AS, 47,* 1.

Hauser, M. G., et al. (1998). The COBE diffuse infrared background experiment search for the cosmic infrared background. *I. Limits and Detections. ApJ, 508,* 25–43.

Högbom, J. A. (1974). Aperture synthesis with a non-regular distribution of interferometer baselines. *A & AS, 15,* 417.

Jonas, J. L., Baart, E. E., & Nicolson, G. D. (1998). The Rhodes/HartRAO 2326-MHz radio continuum survey. *MNRAS, 297,* 977–989.

Laureijs, R. J. et al. (1995). Moderate-Density regions in the LYNDS 134 cloud complex. *ApJS, 101*, 87.

Leitch, E. M. et al. (1997). An anomalous component of galactic emission. *ApJ, 486*, L23+.

Lynds, B. T. (1962). Catalogue of Dark Nebulae. *ApJS, 7*, 1.

Markwardt, C. B. (2009). Non-linear least-squares fitting in IDL with MPFIT. In D. A. Bohlender, D. Durand & P. Dowler (Eds.), Astronomical data analysis software and systems XVIII, Vol. 411. Astronomical Society of the Pacific Conference Series, p. 251.

Mathis, J. S., Mezger, P. G., & Panagia, N. (1983). Interstellar radiation field and dust temperatures in the diffuse interstellar matter and in giant molecular clouds. *A & A, 128*, 212–229.

Mattila, K., Juvela, M., & Lehtinen, K. (2007). Galactic dust clouds are shining in Scattered Hα Light. *ApJ, 654*, L131–L134.

Mattila, K., & Sandell, G. (1979). Observations of neutral hydrogen and OH in the dark nebula LYNDS 1778/1780. *A & A, 78*, 264–274.

Padin, S., et al. (2002). The cosmic background imager. *PASP, 114*, 83–97.

Planck Collaboration et al. (2011). Planck early results. XXV. Thermal dust in nearby molecular clouds. *A & A, 536*, A25.

Planck Collaboration et al. (2013a). Planck 2013 results. I. Overview of products and scientific results. ArXiv e-prints.

Planck Collaboration et al. (2013b). Planck 2013 results. XII. Component separation. ArXiv e-prints.

Planck Collaboration et al. (2013c). Planck 2013 results. XIII. Galactic CO emission. ArXiv e-prints.

Planck Collaboration et al. (2013d). Planck 2013 results. XXXI. All-sky model of thermal dust emission. ArXiv e-prints.

Planck Collaboration et al. (2013e). Planck intermediate results. XV. A study of anomalous microwave emission in Galactic clouds. ArXiv e-prints.

Reich, P., & Reich, W. (1986). A radio continuum survey of the northern sky at 1420 MHz. *II. A & AS, 63*, 205–288.

Reich, P., Testori, J. C., & W. Reich. (2001). A radio continuumsurvey of the southern sky at 1420 MHz. *The atlas of contour maps. A & A, 376*, 861–877.

Reich, W. (1982). A radio continuum survey of the northern sky at 1420 MHz. *I. A & AS, 48*, 219–297.

Ridderstad, M., et al. (2006). Properties of dust in the high-latitude translucent cloud L1780. I. Spatially distinct dust populations and increased dust emissivity from ISO observations. *A & A, 451*, 961–971.

Sault, R. J., Teuben, P. J. & Wright, M. C. H. (1995). A retrospective view of MIRIAD. In R. A. Shaw, H. E. Payne & J. J. E. Hayes (Eds.), Astronomical Data Analysis Software and Systems IV, Vol. 77. Astronomical Society of the Pacific Conference Series, p. 433.

Schlegel, D. J., Finkbeiner, D. P., Davis, M. (1998). Maps of Dust Infrared Emission for Use in Estimation of Reddening and Cosmic Microwave Background Radiation Foregrounds. *ApJ, 500*, 525–+.

Silsbee, K., Ali-Haïmoud, Y., & Hirata, C. M. (2011). Spinning dust emission: the effect of rotation around a non-principal axis. *MNRAS, 411*, 2750–2769.

Snow, T. P., & McCall, B. J. (2006). Diffuse automic and molecular clouds. *ARA & A, 44*, 367–414.

Taylor, A. C., et al. (2011). The cosmic background imager 2. *MNRAS, 418*, 2720–2729.

Toth, L. V. et al. (1995). L 1780: a cometary globule associated with Loop I? *A & A, 295*, 755.

Vidal, M., et al. (2011). Dust-correlated cm wavelength continuum emission from translucent clouds ζ Oph and LDN 1780. *MNRAS, 414*, 2424–2435.

Weingartner, J. C., & Draine, B. T. (2001). Dust grain-size distributions and extinction in the Milky Way, Large Magellanic Cloud, and Small Magellanic Cloud. *ApJ, 548*, 296–309.

Witt, A. N. et al. (2010). On the origins of the high-latitude hα background. *ApJ, 724*, 1551–1560.

Ysard, N., Miville-Deschênes, M. A., & Verstraete, L. (2010). Probing the origin of the microwave anomalous foreground. *A & A, 509*(L1), L1.

Chapter 6
Conclusions and Future Work

In this thesis we have studied the Galactic emission in the 23–43 GHz frequency range, focusing on the diffuse polarised emission and on the anomalous microwave emission. This chapter summarises the results and describes the future work that can be undertaken as a continuation of this research.

6.1 Conclusions

With the aim of obtaining with precision the cosmological parameters from the CMB, great effort is currently being expended on measuring the polarised sky in the 1–1000 GHz frequency range. The detection of primordial B-modes in the CMB spectrum would confirm the inflationary paradigm and would give insights into the extremely high energy scale of the early Universe. Additionally, the quantification of the E-mode spectrum provides a complementary data set which can break degeneracies in cosmological parameters.

One of the main difficulties in these studies is that the polarised sky is dominated by Galactic emission on large angular scales. Below ∼60 GHz, synchrotron radiation dominates the spectrum while thermal dust emission is the main polarised source at higher frequencies. Polarised emission gives information about the Galactic magnetic field and is a fundamental tool to quantify its properties.

6.1.1 Polarisation De-Bias Method

A well known problem while working with polarisation data is the positive bias that affects the quantity $P = \sqrt{Q^2 + U^2}$, and is particularly important at low SNR. Here we worked on the implementation of a de-biasing method based on the knowledge of the *true* polarisation angle χ. At *WMAP* frequencies, we showed that Faraday

© Springer International Publishing Switzerland 2016
M. Vidal Navarro, *Diffuse Radio Foregrounds*, Springer Theses,
DOI 10.1007/978-3-319-26263-5_6

rotation is very small ($\lesssim 1°$ for most of the sky), so the observed polarisation angle χ can be safely taken as the true polarisation angle of the emission.

We corrected the *WMAP* polarisation amplitude maps for the polarisation bias. Using these corrected maps, we set upper limits for the polarisation fraction of two regions where AME dominates between 23 and 40 GHz, the Perseus and ρ Ophiuchi molecular clouds. We found that the polarisation fraction of AME in the ρ Ophiuchi cloud is lower than 1.7, 1.6 and 2.6 % at 23, 33 and 41 GHz respectively. In the Perseus complex, the upper limits are 1.4, 1.9 and 4.7 % at 23, 33 and 41 GHz respectively. If these small values for the AME polarisation are representative over the entire sky, the analysis of the polarisation of the CMB would be simpler, due to the little contribution of this dust-correlated foreground. The results of this work are reported in Dickinson et al. (2011).

6.1.2   WMAP *Polarised Filaments*

The polarised sky at 23 GHz is dominated by synchrotron emission and, away from the Galactic plane, it originates mostly from filamentary structures with well ordered magnetic fields. Some of these structures have been known for decades in radio continuum maps: the "radio loops", with the North Polar Spur being the most studied. The origin of these filaments is not clear and there are many filaments that are visible for the first time in these polarisation data. We have identified 11 filaments, including three of the well known radio continuum loops. 5 of these filaments are only visible in the polarisation data. The geometry of these filaments can be described using circular arcs. We fitted for the centres and radii of these "loops". We also compared polarisation angles along the filaments with the direction defined by their extension. The polarisation angle is well aligned along the filaments, being typically tangential to their direction. We found however some systematic differences between the polarisation angle χ and the direction defined by the extension of the filament. We also measured the polarisation spectral indices of the filaments and 18 smaller regions in the sky. Some of these regions show a spectral index flatter than the usual $\beta = -3.0$. Particularly, some areas at the Fan region, around Galactic longitude $l = 140$, with $\beta_{K-Ka} = -2.68 \pm 0.16$. The average spectral index in all the regions considered is $\beta_{K-Ka} = -3.04 \pm 0.02$. On average, no significant steepening of the spectral index is observed between β_{K-Ka} and β_{Ka-Q}, although some individual regions do show some steepening. We found significant variations in β over the sky.

We showed that the observed Faraday rotation is very low. For most of the sky, the variation in the polarisation angle ($\Delta\chi$) is less than $1°$ at K–band and at higher frequencies. The Galactic centre region shows the largest Faraday rotation. Here we found a rotation measure $RM = -4382 \pm 204 \, \mathrm{rad \, m^{-2}}$, close to previous values from the literature.

We also quantified the polarisation fraction of the synchrotron emission over the entire sky and in a narrow region of the Galactic plane. We used different templates to estimate the synchrotron total intensity at 23 GHz. Some of the polarised filaments show a large polarisation fraction ($\Pi \approx 40\,\%$) regardless of the synchrotron total intensity template used. On the Galactic plane, in the region $20° < l < 44°$, we can make a better estimation of the polarisation fraction using an accurate free-free map, which was prepared using hydrogen RRL by Alves et al. (2012). By subtracting the free-free contribution from the 0.408 GHz map from Haslam et al. (1982), we produce a pure synchrotron template. Scaling this template up to 23 GHz, we found that the polarisation fraction has a peak value of $\Pi = 16.5\,\%$. We also show that the polarisation amplitude on the plane can be described by a narrow Gaussian component, with a FWHM = $1°.9$. The total intensity of the synchrotron emission shows a similar narrow component with FWHM = $1°.8$, in addition to a broader component which we describe using a parabolic curve.

To explain the large-scale observed polarisation pattern, we invoke a model originally proposed by Heiles (1998), in which an expanding shell, located at a distance of 120 pc compresses the magnetic field in the local ISM. Under the assumption that the unperturbed magnetic field lines are parallel to the Galactic plane, an expanding spherical shell will bend the lines in a simple manner. We calculated how these field lines would appear from our vantage point. The observed polarisation angles are in good agreement with the direction of the field lines for most of the relevant area of the sky.

Finally, we estimated the level of contamination that the diffuse filaments add to the CMB E- and B-mode power spectra. We compared the B-mode power spectrum of the sky using two different masks, one that covers only the Galactic plane and a second one that masks-out the diffuse filaments. The power measured at $\ell = 3$ using the Galactic plane masks is $\sim$140 times larger than the power measured using the filaments mask. This implies that a similar masking of the filaments, or alternatively, a careful subtraction is required to precisely measure the CMB spectrum at the largest angular scales.

6.1.3 QUIET Galactic Maps

The Q/U Imaging ExpetimenT observed the polarised sky at 43 and 95 GHz with the aim of measure the CMB E-mode power spectrum and set competitive limits on the B-mode spectrum. It uses two arrays of 17 and 84 polarimeters for Q–band and W-band respectively, based on High Electron Mobility Transistors (HEMT). They can measure simultaneously both Stokes Q and U parameters. These two arrays are the largest HEMT-based arrays ever built. The 43 GHz array, has the best sensitivity ($69\,\mu K\sqrt{s}$) and the lowest instrumental systematic errors ever achieved in this band, which contribute to setting limits on the tensor-to-scalar ratio of $r < 0.1$. The W-band array, also presents the lowest systematic error to-date and it contributes to r at the $r < 0.01$ level.

Observing from an altitude of 5080 m in the Atacama desert, QUIET observed 6 regions of $15° \times 15°$: 4 CMB patches and 2 regions on the Galactic plane. From the CMB observations, a significant detection is made of the E-mode power spectrum, and the results agree with the ΛCDM model. Upper limits are set for the B-mode spectrum, limiting $r < 2.7$ at 95 % C.L.

Synchrotron emission was statistically detected in one of the CMB fields in the 43 GHz data. We tested the level of foreground contamination of the 95 GHz maps from this field by measuring the distribution of the polarisation angles on the map. A pure CMB map should have random polarisation angles. We showed, after proper filtering of the map, that the distribution of angles is Gaussian as expected, so the foreground level is below the statistical error in this field.

We also worked on the definition of the filtering used in the map-making of the Galactic maps at 43 GHz. This filtering is different from the one used in the CMB fields, due to the high SNR on the Galactic plane. The produced maps are consistent with the *WMAP* Q–band polarisation data, and show a better SNR. The filtering of the QUIET maps, however, makes a direct comparison with the *WMAP* data difficult. Nevertheless, the spectral indices measured between *WMAP* K–band and QUIET Q–band are consistent with those measured using only *WMAP* data. We also found a systematic difference in the polarisation angles between the two Galactic fields observed. This difference can be explained by large-scale (30°) variations in the magnetic field of the Galactic plane.

6.1.4   AME in LDN 1780

Anomalous microwave emission is observed in the $\sim 10 - 50$ GHz frequency range, and is thought to be originated by small dust grains spinning at GHz frequencies. AME has been observed in different astrophysical environments, dust clouds, HIIregions, and statistically in the diffuse cirrus clouds at high Galactic latitudes.

Following the detection of AME in the LDN 1780 translucent cloud presented in Vidal et al. (2011) using data from the Cosmic Background Imager (CBI), we performed follow-up observations at 31 GHz using the CARMA 3.5-m array. These new data have an angular resolution of $2'$, 3 times better than that of the CBI. We measured the correlation between the 31 GHz data and different IR templates. We found that the best correlation occurs with MIPS 70 μm, with a Spearman's rank $r_S = 0.49 \pm 0.07$. This confirms what we found using the CBI data in Vidal et al. (2011), where the radio data correlated better with a 60 μm IR map. The correlation between the CARMA data and *Spitzer* 8 μm and *Spitzer* 24 μm, which traces PAHs and VSGs, is very poor. Here, $r_S = 0.14 \pm 0.06$ and $r_S = 0.21 \pm 0.06$ for 8 and 24 μm respectively. These two correlation values increase significantly when correcting the IR maps by the ISRF, yielding $r_S = 0.38 \pm 0.07$ and $r_S = 0.46 \pm 0.06$ for the corrected 8 and 24 μm templates respectively. This is important as these ISRF corrected templates should be better tracers of the PAHs and VSGs column density, as opposed to the uncorrected maps, in which the emission is proportional also to the incoming radiation field.

We constructed an SED of LDN 1780 on $1°$ scales between 0.408 and 2997 GHz, including *Planck* data. The region of the cloud is dominated by the CMB anisotropy between 23 and 217 GHz, but it is clear also the presence of AME, which is well fitted using a spinning dust model.

On the 31 GHz CARMA maps, there are differences in the emissivity along the cloud. These differences are compatible with the spinning dust model. The spinning dust emission depends on the physical parameters of the dust grain and also on environmental conditions of the cloud, such as density, temperature and ISRF. Some of these parameters for LDN 1780 are known from the literature, so we fixed them and concluded that variations in the grain size distribution along the cloud can reproduce an observed factor $\sim$6 difference in the AME emissivity.

Given the large number of free parameters that the spinning dust models have, it is not difficult to account for AME variations in different environments. This is something to keep in mind when trying to interpret the observations. It is clear that a greater number of radio-observations are required in order to fully constrain the spinning dust models and test if this is really the emission mechanism of the AME.

6.2 Future Work

There are a number of things than can be done in the different topics studied. Here we describe what we will like to pursuit regarding the subject of this thesis.

6.2.1 Polarised Diffuse Emission

The modelling of the large-scale polarisation filaments seen in *WMAP* data is an interesting topic. It is still not clear what the origin is of the large filaments. The lack of a good measurement of the distance to these objects is the main complication to explain their nature. We have shown that the simple model from Heiles (1998) explains reasonably well the distribution of the polarisation angles of the largest filaments. A more detailed modelling, including predictions for the variations of the synchrotron intensity across the sky, will help to affirm the hypothesis of a local (a few hundred pc) origin of the large filaments. The emission from the NPS region is much larger, ($\geq$500 times) than the mean Galactic synchrotron emissivity around the Sun (Heiles 1998). This difference in brightness is something that a complete model should provide. This kind of modelling should ideally include the 3D density distribution of the ISM in the solar neighbourhood.

There is evidence that some of the filaments lie at distances larger than a few kilo-parsecs, as the results presented by Carretti et al. (2013) and Sun et al. (2014). There are also some thin filaments that trace very well the border of the "*Fermi* bubbles". Here, it is difficult to neglect the connection between the polarised emission and the gamma-ray data. New data that is soon to be released, such as the 5 GHz polarisation maps from the CBASS collaboration and the polarisation maps from *Planck* will greatly help.

Another very interesting aspect that can help in the determination of the distance to some of the polarised structures is the discovery that we made of an anti-correlation between Hα emission and the polarised intensity at *WMAP* K–band . There are a few thin filaments visible on the Hα maps that have a correspondence with the polarisation amplitude. In Fig. 6.1 we show on *top* an Hα map, integrated in the velocity range $-80 < v_{lsr} < -40\,\mathrm{km\,s^{-1}}$ from the Wisconsin H-Alpha Mapper (WHAM) survey (Haffner et al. 2003). The filament encircled within the black ellipse is ∼40° in length. The plot in the middle corresponds to the *WMAP* 23 GHz polarised intensity, which shows a trough at the same location as the Hα filament. This feature is also visible on the Faraday depth map from Oppermann et al. (2012), shown at the *bottom* of the same figure.

The origin of this anti-correlation between Hα intensity and polarisation amplitude is not clear. A first possibility is that the trough visible in the *WMAP* polarisation map is a depolarised region, meaning that the filament Hα filament lies in between the polarised background emission and us. If this is the case, the ionised gas traced by the Hα map produces Faraday rotation, due to the presence of a magnetic field in the plasma, depolarising the diffuse background emission along its extension. This hypothesis however is not compatible with the low density of the ionised gas. The intensity of the filament in Hα is 2 R above the background at the original resolution (6 arcmin) of the Finkbeiner (2003) Hα map. This corresponds to an electron density of $2.1\,\mathrm{cm^{-3}}$, which is very low to produce a considerable Faraday rotation at 23 GHz for typical values of the magnetic field. Moreover, the Faraday depth map from Oppermann et al. (2012) has a value of ∼60 rad m^{-2} along the filament, which corresponds to a 0°.6 change in polarisation angle at 23 GHz. Therefore, Faraday rotation of this high latitude filament is not enough to cause any major depolarisation at 23 GHz.

A second alternative is that the observed polarisation feature is intrinsic, i.e. there is no synchrotron emission emanating from the region of the Hα filament. This might be due to a less organised magnetic field in this region in comparison with the diffuse emission seen in the vicinity of the filament on the polarisation map.

We note also the fact that the Hα filament is only visible at negative radial velocities. This corresponds to the same velocity range of the Perseus arm of the Galaxy. If the Hα filament belongs to that arm, it will imply that the distance to the diffuse synchrotron background is much larger than a few hundred parsecs, lying at least 2 kpc away from us.

Even more interesting, this feature is not unique. A similar anti-correlation can be observed next to the NPS, also in Fig. 6.1. A dim filament running from $(l_0, b_0) \approx (30°, 30°)$ to $(l_1, b_1) \approx (10°, 60°)$ is visible on the Hα map. The polarisation map also presents emission here. In this case, the Faraday depth map do not show any feature with the same morphology. A detailed observation of the maps reveals more of these structures.

The fact that we see several features showing similar anti-correlation, might suggest a similar explanation for all of them. A detailed study is needed, adding the *Planck* polarisation data, which should be available in 2014, will increase the SNR.

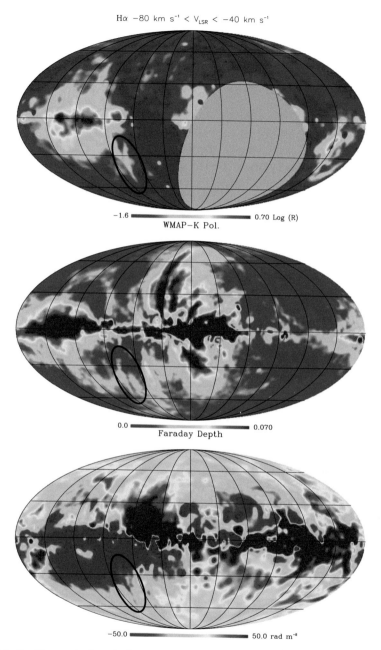

Fig. 6.1 *Top* Hα map in the velocity range $-80 < v_{lsr} < -40\,\mathrm{km\,s^{-1}}$. Note the vertical filament that runs at $l \approx 75°$ and $-60° b < -20°$. This filament has a counterpart in "absorption", visible as a trough in the *WMAP* K–band polarisation amplitude map in the *middle*. The Faraday depth map from (Oppermann et al. 2012) at the *bottom* also shows a filament at the same location, with a mean value of $-50\,\mathrm{rad\,m^{-2}}$ along its extension. The three maps have a common resolution of 3° and the grid spacing is 30°

Also, the CBASS polarisation survey at 5 GHz data will be useful to constrain the level of Faraday depolarisation on these objects.

6.2.2 AME

The spinning dust emissivity depends on a number of parameters which affect the emission in a complicated manner. We will perform a more detailed modelling, allowing the parameters to vary over a grid of values, instead of fixing some of them as we have done in this work. This will allow us to understand the real dependence of the spectrum with respect to the different possible combinations of parameters and from this, draw probability distribution for the preferred values that reproduce the observed emissivities.

The spinning dust model, in general, predicts a peak frequency that is higher than the 31 GHz at which we have observed LDN 1780. Observing the cloud at a frequency $\gtrsim$40 GHz we should measure a larger AME brightness. We know that there are variations in AME emissivity along the cloud which are related to the dust properties. We would like to observe two or three regions of the cloud, at two difference frequencies, to measure the spectral index of AME. This would be a good test for the spinning dust hypothesis as we can predict before hand the spectral indices on the different regions using the SPDUST code. Deep observations using a sensitive instrument like the Green Bank Telescope will be ideal for this.

Regarding AME polarisation, the soon-to-be-released *Planck* data, in conjunction with *WMAP* data, will increase the SNR of the polarised maps of the sky in the 30–90 GHz. There are a number of clouds, particularly at high Galactic latitudes that are good targets to test any polarisation from AME. Furthermore, we could constrain even further the limits that exist already for the ρ Ophiuchi and Perseus regions. Hopefully it might be possible to detect AME polarisation. Such detection would be of great use in the understanding of this emission mechanism, as different AME models predict distinct values for its polarisation. Ultimately, AME could be used as an additional test for the physics of the ISM.

References

Alves, M. I. R. et al. (2012). A derivation of the free-free emission on the Galactic plane between l = 20°. In *MNRAS* (Vol. 422, pp. 2429–2443).

Carretti, E. et al. (2013). Giant magnetized outflows from the centre of the milky way. In *Nature* (Vol. 493, pp. 66–69).

Dickinson, C., Peel, M., & Vidal, M. (2011). New constraints on the polarization of anomalous microwave emission in nearby molecular clouds. In: *MNRAS* (Vol. 418, pp. L35–L39).

Finkbeiner, D. P. (2003). A full-sky Hα template for microwave foreground prediction. In *ApJS* (Vol. 146, pp. 407–415).

Haffner, L. M. et al. (2003). The Wisconsin Hα mapper northern sky survey. In *ApJS* (Vol. 149, pp. 405–422).

Haslam, C. G. T. et al. (1982). A 408 MHz all-sky continuum survey. II—The atlas of contour maps. In *A & AS* (Vol. 47, p. 1).

Heiles, C. (1998). The magnetic field near the local bubble. In D. Breitschwerdt, M. J. Freyberg, & J. Truemper (Eds.), *IAU Colloq. 166: The Local Bubble and Beyond* (Vol. 506, pp. 229–238). Lecture Notes in Physics, Berlin: Springer.

Oppermann, N., et al. (2012). An improved map of the galactic Faraday sky. *A & A, 542*(A93), A93.

Sun, X. H., et al. (2014). Absolutely calibrated radio polarimetry of the inner Galaxy at 2.3 and 4.8 GHz. *MNRAS, 437*, 2936–2947.

Vidal, M., et al. (2011). Dust-correlated cm wavelength continuum emission from translucent clouds ζ Oph and LDN 1780. *MNRAS, 414*, 2424–2435.

Index

© Springer International Publishing Switzerland 2016
M. Vidal Navarro, *Diffuse Radio Foregrounds*, Springer Theses,
DOI 10.1007/978-3-319-26263-5